the STUFF of LIFE

Written by MARK SCHULTZ

Art by ZANDER CANNON and KEVIN CANNON

A Novel Graphic from Hill and Wang

A division of Farrar, Straus and Giroux

New York

Hill and Wang
A division of Farrar, Straus and Giroux
18 West 18th Street, New York 10011

This is a Z FILE, INC. Book

Distributed in Canada by Douglas & McIntyre Ltd.
Printed in the United States of America
Published simultaneously in hardcover and paperback
First edition, 2009

Library of Congress Cataloging-in-Publication Data

Schultz, Mark, 1955-
The stuff of life : a graphic guide to genetics and DNA / written by Mark Schultz ; art by Zander Cannon and Kevin Cannon.— 1st ed.
p. cm
Includes bibliographical references and index.
ISBN-13: 978-0-8090-9494-3 (hardcover : alk. paper)
ISBN-10: 0-8090-9494-0 (hardcover : alk. paper)
ISBN-13: 978-0-8090-8938-3 (pbk. : alk. paper)
ISBN-10: 0-8090-8938-6 (pbk. : alk. paper)
1. Genetics—Comic books, strips, etc. 2. DNA—Comic books, strips, etc. I. Cannon, Zander. II. Cannon, Kevin. III. Title.

QH437.S348 2009
576.5—dc22

2008033829

Editor: Howard Zimmerman
Design: Zander Cannon and Kevin Cannon
Science Consultant: Dave C. Bates
Production Assistant: Elizabeth Maples

www.fsgbooks.com

1 3 5 7 9 10 8 6 4 2

For Grace

--Mark Schultz

This book, and every book forever and ever, is for Julie

--Zander Cannon

For R.A.

--Kevin Cannon

The Stuff of Life represented a significant step into new territories for me. As such, it required an unusually high degree of consideration and cooperation among all those involved. My thanks to Zander Cannon and Kevin Cannon, who took the script and made it much more than the sum of description and text; to Dave Bates, our scientific consultant, who astutely kept us on the razor's edge between factual correctness and format necessity; to Howard Zimmerman, who conceived this project and wrangled it through, wielding a deft and thoughtful editorial hand; and to Thomas LeBien, whose vision and open-mindedness made it all possible.

--Mark Schultz

We would like to thank the many generous individuals who have helped make this book a reality. Everyone involved in this project has been extremely enthusiastic and exceedingly helpful to us from day one. Thanks to Mark Schultz for handing us a script that was truly a pleasure to draw, to David Bates for his knowledge and commitment to making sure we got our science facts straight, and to Howard Zimmerman for being our fantastic editor, sounding board, liaison, and cheerleader. We all worked hundreds--sometimes thousands--of miles apart, but it never felt that way. Finally, a huge thanks to Thomas LeBien at Hill and Wang, the man responsible for realizing this gorgeous book and for taking a chance on this unique way of teaching science.

--Zander Cannon and Kevin Cannon

CONTENTS

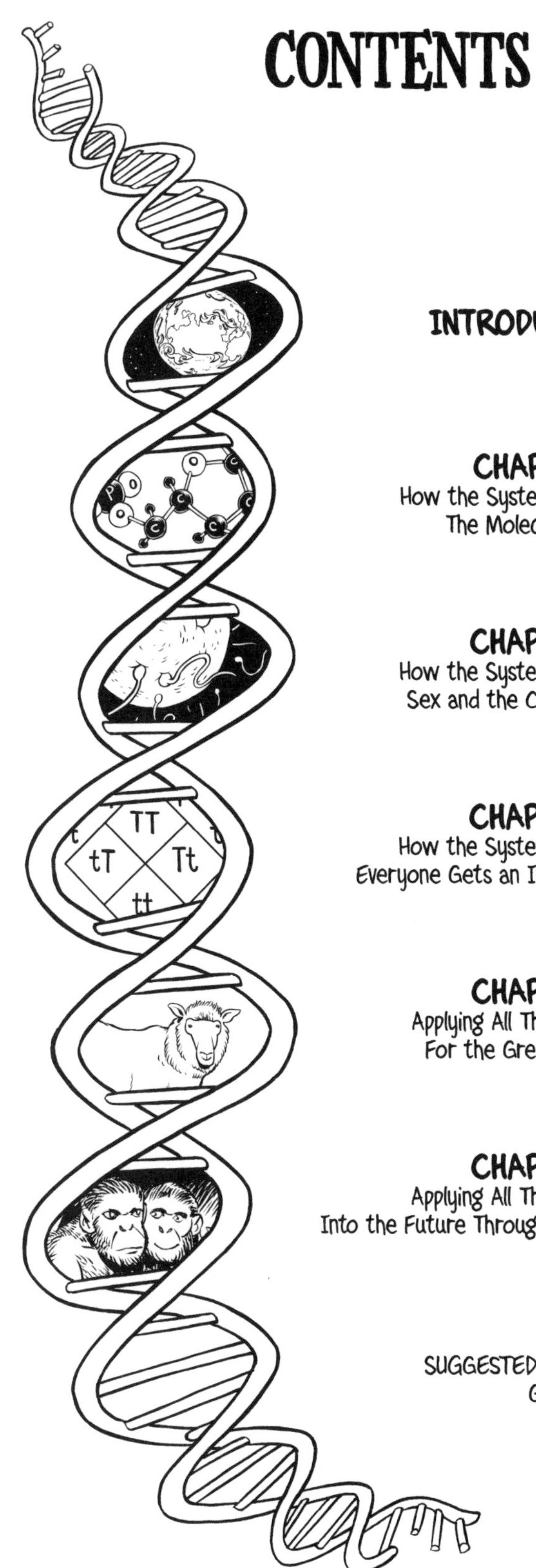

the STUFF of LIFE

the STUFF of LIFE
A graphic guide to genetics and DNA
WORDS Mark Schultz
Zander Cannon
Kevin Cannon ART
A DETAILED TRANSLATION OF THE REPORT OF BLOORT 183, INTERPLANETARY BIOLOGIST OF THE GLARGAL ROYAL SCIENCE ACADEMY, TO HIS SUPREME HIGHNESS FLOORSH 727, CONCERNING THE UNIVERSAL REGENERATIVE AND REPRODUCTIVE STRATEGY OF LIFE ON THE PLANET EARTH, SATELLITE OF SOL.
AN EDIFYING AND SOMETIMES SHOCKING EXPLORATION OF THE GREAT DIVERSITY, TENACITY, AND INVENTIVENESS OF LIFE ON THAT WORLD, AND A CRITICAL LOOK AT THE DOMINANT SPECIES' EVOLVING MASTERY OF GENETIC MANIPULATION.
FEATURING ASSORTED VISUAL AIDS AND SIDEBARS.

THIS IS EARTH, AS IT APPEARS TODAY.
IT'S A MEDIUM-SIZE PLANET WITH ENOUGH MASS TO HOLD A VIABLE ATMOSPHERE, BUT NOT SO MUCH AS TO CRUSH ASPIRATIONS.
EARTH'S EARLY HISTORY WAS FAIRLY TYPICAL OF SUCH CELESTIAL BODIES: FIVE BILLION YEARS AGO IT WAS NO MORE THAN A RIVER OF DUST AND ROCKY FRAGMENTS CIRCLING SOL, ITS SUN.
OVER ENOUGH TIME, EVEN DUST CLOTTING TOGETHER CAN GAIN ENOUGH MASS TO CREATE A GRAVITY CHAIN REACTION.
4.6 BILLION YEARS AGO
THROUGH THE ACTIONS OF GRAVITY AND ACCRETION, EARTH WAS BORN--
--A SEETHING, MOLTEN WORLD COALESCING UNDER THE IMPACTS OF LARGER AND LARGER ADDITIONS.
THE RADIOACTIVE DECAY OF ISOTOPES DEEP WITHIN THIS TORTURED BALL CREATED AND STOKED A CONSTANT FURNACE AT ITS CORE.
4.3 BILLION YEARS AGO
THE SURFACE WAS CONTINUALLY PELTED WITH A RAIN OF METEORITES BEARING EXOTIC ELEMENTS-- MOST NOTABLY CARBON, AS WE'LL COME TO SEE.
THEN A LONG PERIOD OF COOLING BEGAN, WITH THE YOUNG PLANET'S BOWELS DISGORGING THE GASES AND STEAM THAT CONTRIBUTED TO ITS OCEANS AND ITS NASCENT ATMOSPHERE--
4.1 BILLION YEARS AGO
--SNUGLY HELD IN PLACE BY EARTH'S GRAVITY-GENERATING MASS.
ALL THESE FACTORS, PLUS EARTH'S CONGENIAL DISTANCE FROM SOL, ITS ROTATION, AND THE SUBTLE TILT OF AN AXIS THAT DISPERSES SOLAR HEAT AROUND ITS SURFACE...
...CREATED THE PERFECT CONDITIONS FOR THE EXTRAORDINARY PHENOMENON KNOWN AS...

LIFE!
3.9 BILLION YEARS AGO
THE TRANSMUTATION OF BASE ELEMENTS INTO COMPOUNDS JUMP-STARTED BY A SUPPLY OF ENERGY INTO SELF-PERPETUATING ORGANISMS! THE CAPACITY TO GROW, TO REACT TO STIMULI, TO REPLICATE, RENEW, AND REPRODUCE!
EARTH'S FIRST LIFE WAS PROBABLY AKIN TO THE SULFUR-EATING HYPERTHERMOPHILE BACTERIA THAT STILL EKE OUT THEIR EXISTENCE IN THE MOST EXTREME ENVIRONMENTS, DOING QUITE WELL IN THEIR GEOLOGICALLY ACTIVE NICHES WITHOUT THE BENEFIT OF SUNLIGHT OR OXYGEN.
BUT THE ORGANISMS THAT POINTED TO THE FUTURE OF LIFE ON EARTH WERE THE CYANOBACTERIA THAT INVENTED PHOTOSYNTHESIS-- THE ABILITY TO CREATE THEIR OWN ENERGY FROM SUNLIGHT.
3 BILLION YEARS AGO
FOR MILLIONS OF YEARS VAST COLONIES OF THEM-- MATTED AND STACKED INTO STROMATOLITES-- DOMINATED THE PLANET...
...AND, AS A SIDE PRODUCT OF THEIR PRODUCTION OF ENERGY FROM WATER AND SUNLIGHT, RELEASED INCREDIBLY HUGE QUANTITIES OF OXYGEN, AIDING THE FORMATION OF A NEW ATMOSPHERIC MIX--
2.5 BILLION YEARS AGO
--ONE POISONOUS TO MANY OF THE EARLIER ORGANISMS...
...BUT OFFERING FANTASTIC OPPORTUNITIES FOR BOLD, NEW EXPERIMENTATION. SOME NOVEL PROTIST ENVELOPED AND CAPTURED BACTERIA, AND THEN FOUND THAT THEY ALL COULD WORK TOGETHER-- SYMBIOSIS!
2.1 BILLION YEARS AGO
THE BACTERIA EVOLVED INTO COOPERATIVE ORGANELLES, AND COMPLEX CELLS WERE BORN!
COMPLEX CELLS BEGAN JOINING TO FORM MULTI-CELLULAR ORGANISMS. AT SOME POINT NOW LOST IN TIME, SOME OF THEM EVOLVED A REPRODUCTIVE METHOD CALLED SEX, WHICH ALLOWS FOR THE CHARACTERISTICS OF TWO PARENT ORGANISMS TO BE PASSED IN RICH, VARIABLE COMBINATIONS TO A NEW GENERATION, AND THINGS STARTED EVOLVING MUCH FASTER!
1.2 BILLION YEARS AGO

MULTI-CELLULAR ORGANISMS CONTINUED TO EVOLVE AND DIFFERENTIATE INTO THREE BROAD CLASSIFICATIONS:
PLANT
ANIMAL
FUNGUS
PLANTS ARE AUTOTROPHS -- THEY PRODUCE THEIR OWN FOOD BY PHOTOSYNTHESIS, THE SAME TECHNIQUE PIONEERED BY THE ANCIENT CYANOBACTERIA.
LARGELY IMMOBILE, THEY ARE IN CONSTANT COMPETITION FOR THE BEST LIGHT CONDITIONS.
FUNGI AND ANIMALS ARE HETEROTROPHS -- CONSUMERS OF FOOD SOURCES PRODUCED BY OTHERS.
ANIMALS THEMSELVES ARE FURTHER DIVIDED INTO TWO GROUPS --INVERTEBRATES, WHICH, LIKE THIS LOVELY, DELICATELY FORMED SEA CUCUMBER, ARE NOT ENCUMBERED BY A SPINAL CORD...
...AND VERTEBRATES, WHICH HAVE DEVELOPED A SPINAL CORD. IN SOME CASES, VERTEBRATES HAVE EVEN DEVELOPED ENTIRE INTERNAL SKELETAL SYSTEMS, WHICH COME IN HANDY, I GUESS, IF YOU WANT TO LIVE ON LAND AND ARE OBSESSED WITH SIZE.
ACTUALLY, SOME OF THE MOST INNOVATIVE AND EXCITING RECENT EVOLUTIONARY ADAPTATIONS HAVE BEEN EXPRESSED BY THE WARM-BLOODED VERTEBRATES KNOWN AS BIRDS AND MAMMALS...
...WITH THE MAMMALIAN SUBCLASS CALLED PRIMATES DEVELOPING A SPECIALIZATION IN THAT PARTICULARLY FASCINATING CHARACTERISTIC --
INTELLIGENCE!

SO, HERE WE HAVE THE PRIMATE, MAN, IN HIS CURRENT STATE.
MAN IS THE LONE ORGANISM ON EARTH THAT HAS DEVELOPED A SCIENTIFIC AND TECHNOLOGICAL CULTURE AND IS THUS OF PARTICULAR INTER--
Excuse me...?!
Let me get this straight-- the creature you called the "sea cucumber"?
The sea cucumber ISN'T Earth's dominant organism?
WELL, I WAS CONFUSED AT FIRST, TOO, YOUR ROYAL INQUISITIVENESS. I MEAN-- THE CUKES DO LOOK SOMETHING LIKE US.
BUT THE AMAZING TRUTH IS-- THEY ARE AS DUMB AS HAMMERS.
IN FACT, NO EARTHLY INVERTEBRATE HAS EVER DEVELOPED INTO A GLOBALLY DOMINANT SPECIES, AS WE DID HERE ON GLARGAL.
NO, SIRE, ON EARTH IT WAS THE VERTEBRATES...
...LAND CREATURES PAINFULLY BATTLING GRAVITY WITH WHAT MUST BE TERRIBLY ITCHY INTERNAL SKELETONS...
...THAT EVOLVED INTO THE PLANET'S TOP PREDATOR.
So you're telling me that VERTEBRATES--
--hairy, bipedal VERTEBRATES...
YES. AND THEY ARE NOW DOING REMARKABLE RESEARCH INTO THEIR BIOLOGICAL COMPOSITIONS AND ORIGINS, AS WELL AS INTO--
OKAY--let's... let's hold up a moment, Bloort...
...I need some time to DIGEST this.

I mean-- I never considered...
...I never GUESSED that our counterparts on Earth might be so completely-- ALIEN.
Hairy, bipedal, land-dwelling vertebrates-- HMMPH.
Kush 1290-- are we sure that this... HUMAN SPECIES offers anything WE can use?
AN INSIGHTFUL QUESTION, MOST CAUTIOUS ONE. BUT, AS OUR GENETIC CRISIS HAS BECOME INCREASINGLY DIRE, I BELIEVE WE MUST LOOK TO OTHER SYSTEMS FOR HELP.
YES-- THE INFORMATION WILL BE A BIT CHALLENGING. AS YOUR SCIENCE ADVISER, I URGE YOU TO EXTEND A LITTLE TRUST...
...AND I'M SURE YOU WILL LEARN A GREAT DEAL FROM BLOORT 183. HE'S ONE OF OUR FINEST SCIENTISTS-- AN EXCELLENT OBSERVER AND THOROUGH RESEARCHER.
THERE WAS GOOD REASON WE TRANSMITTED HIM TO EARTH...
I MAKE NO ASSUMPTIONS, O MAGNIFICENT CORPULENCE. ALL MY RESEARCH IS BACKED BY EVIDENCE, PRESENTED HERE THROUGH THE SIMULACRON. AND I DO BELIEVE WE CAN LEARN MUCH FROM THE EARTHLY ORGANISMS...
Of course-- very well. You'll pardon me if my concern over the Royal Family's growing... DIFFICULTIES ...with the HERITABLE DISORDER have made me snappish. Of course I trust Bloort. We expect that what he has learned can only HELP us master our genetic TROUBLES.
It's just, frankly, I'm a bit disappointed. The sea cucumber was so much more RELATABLE.

I THINK, SIRE, THAT IT'S IMPORTANT FOR US TO REALIZE THAT THE SIMILARITY WE SEEM TO SEE BETWEEN THE EARTHLY SEA CUCUMBERS AND OURSELVES IS DECEIVING. IT IS A COINCIDENCE OF SEPARATE EVOLUTIONS MOLDED BY SIMILAR ENVIRONMENTAL FACTORS.
FOR INSTANCE, WE BOTH LIVE IN FLUID ENVIRONMENTS. MORE IMPORTANT, THOUGH: WE COME FROM TWO DIFFERENT PLANETS.
SO, DESPITE VISUAL SIMILARITIES, TWO DIFFERENT PLANETS + TWO DIFFERENT ORIGINS = NO BIOLOGICAL SIMILARITIES WHATSOEVER. ABSOLUTELY NOTHING IN COMMON.
≠
TRUST ME, SIRE. YOU WOULD NOT RELATE WELL TO AN EARTHLY SEA CUCUMBER.
BUT-- TO GET US BACK ON TRACK -- DESPITE THE APPARENT DIFFERENCES BETWEEN SEA CUCUMBERS AND THOSE HAIRY MAMMALIAN BIPEDS--
=
-- THE FACT IS THAT THEY ARE REALLY NOT ALL THAT DIFFERENT!
AT ITS MOST FUNDAMENTAL LEVEL, ALL EXPRESSIONS OF LIFE ON EARTH ARE LINKED VERY CLOSELY TOGETHER. THEY SHARE ONE COMMON ANCESTOR.
AND THE ASTONISHING EVIDENCE LIES IN EVERY CELL THAT EVER LIVED ON THAT PLANET: EVERY ORGANISM OF EARTH, FROM BACTERIA TO HUMANS, SHARES A SINGLE UNIFYING SUBSTANCE THAT DICTATES EVERY ORGANISM'S GROWTH AND REPRODUCTION --
-- A DOUBLE-HELIX SUPER-MOLECULE EARTHLINGS CALL...
DNA!

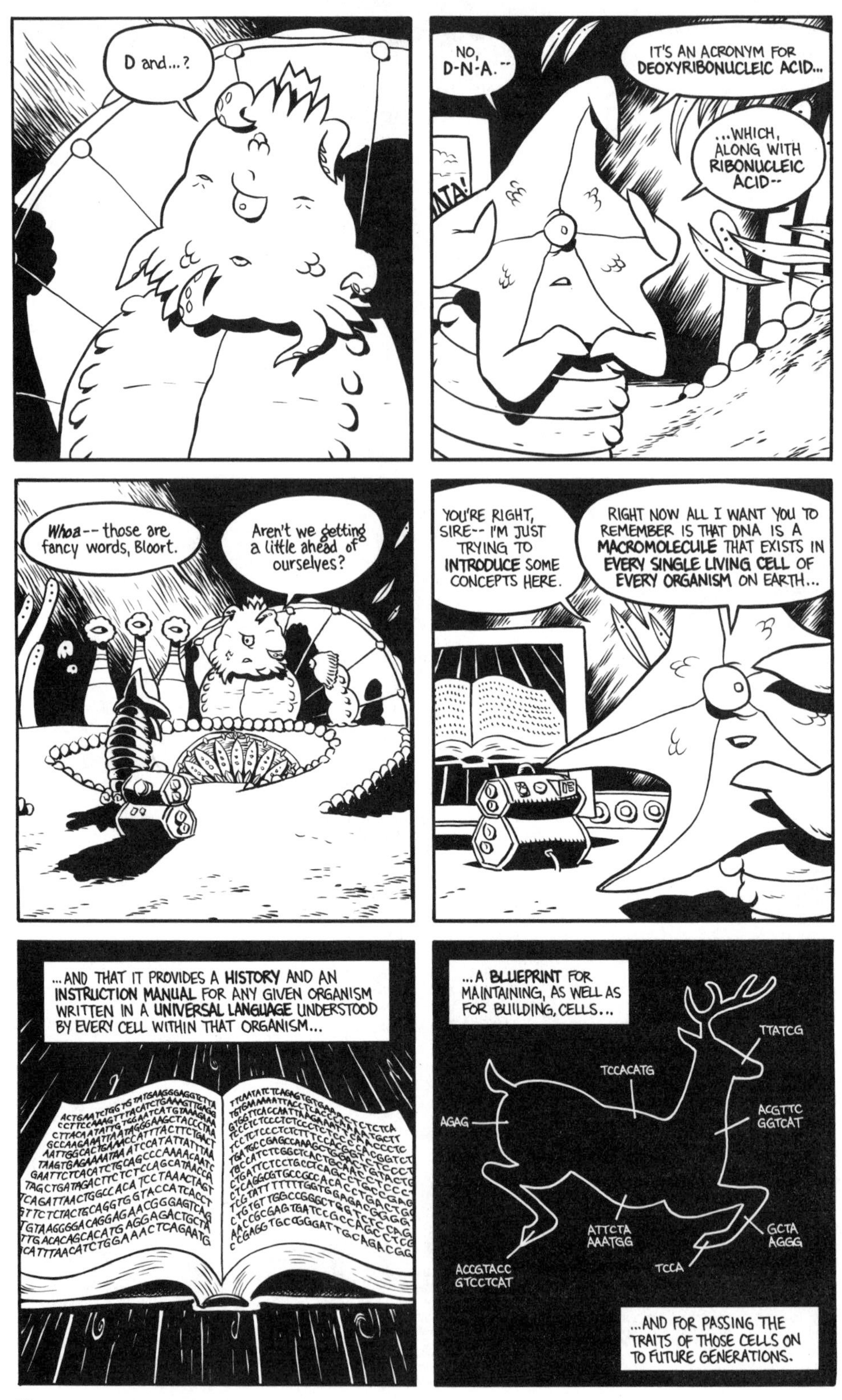
D and...?
NO, D-N-A.--
IT'S AN ACRONYM FOR DEOXYRIBONUCLEIC ACID...
...WHICH, ALONG WITH RIBONUCLEIC ACID--
Whoa-- those are fancy words, Bloort.
Aren't we getting a little ahead of ourselves?
YOU'RE RIGHT, SIRE-- I'M JUST TRYING TO INTRODUCE SOME CONCEPTS HERE.
RIGHT NOW ALL I WANT YOU TO REMEMBER IS THAT DNA IS A MACROMOLECULE THAT EXISTS IN EVERY SINGLE LIVING CELL OF EVERY ORGANISM ON EARTH...
...AND THAT IT PROVIDES A HISTORY AND AN INSTRUCTION MANUAL FOR ANY GIVEN ORGANISM WRITTEN IN A UNIVERSAL LANGUAGE UNDERSTOOD BY EVERY CELL WITHIN THAT ORGANISM...
...A BLUEPRINT FOR MAINTAINING, AS WELL AS FOR BUILDING, CELLS...
TTATCG
TCCACATG
AGAG
ACGTTC
GGTCAT
ATTCTA
AAATGG
GCTA
AGGG
ACCGTACC
GTCCTCAT
TCCA
...AND FOR PASSING THE TRAITS OF THOSE CELLS ON TO FUTURE GENERATIONS.

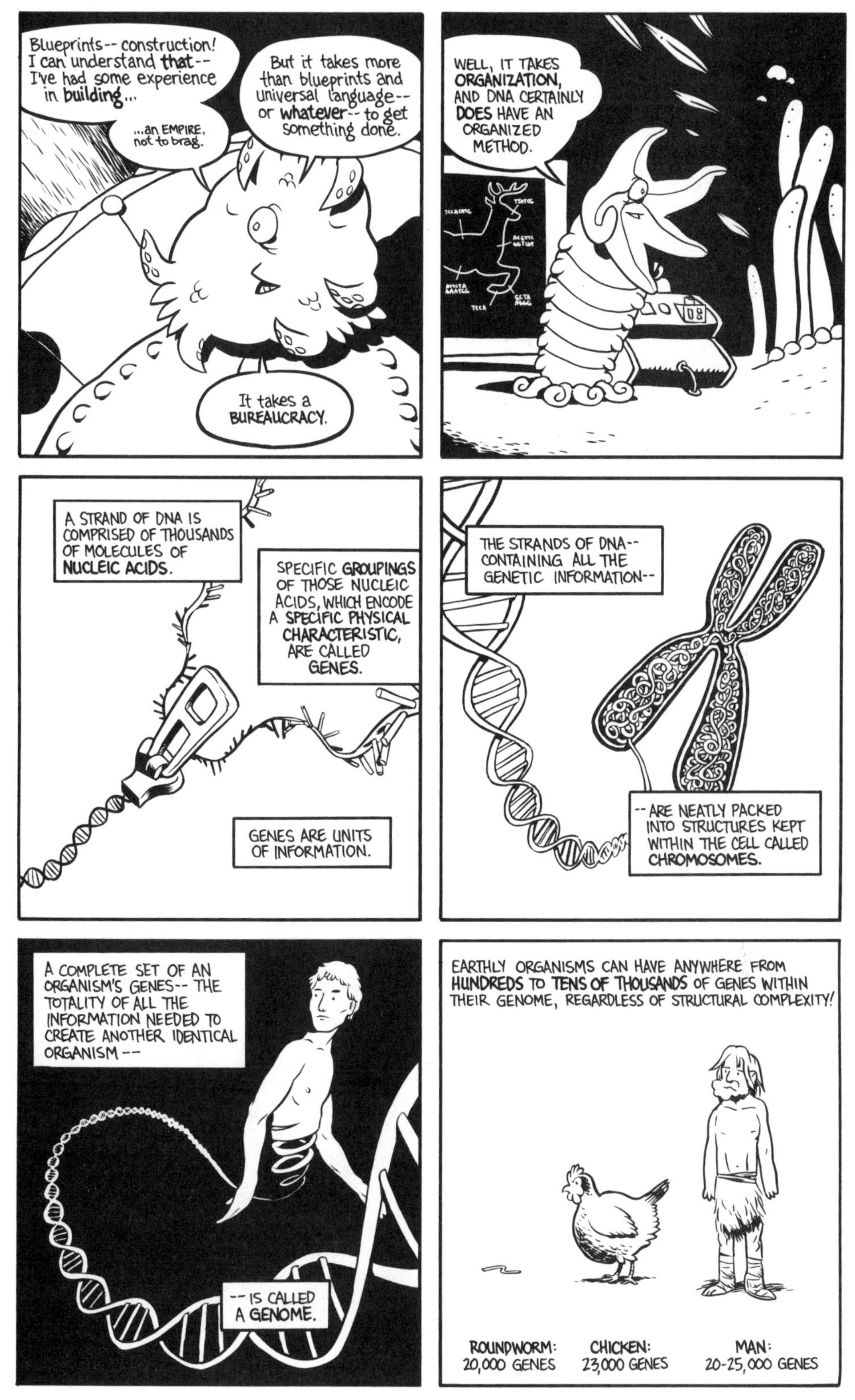
Blueprints-- construction! I can understand that-- I've had some experience in building...
...an EMPIRE, not to brag.
But it takes more than blueprints and universal language-- or whatever-- to get something done.
It takes a BUREAUCRACY.
WELL, IT TAKES ORGANIZATION, AND DNA CERTAINLY DOES HAVE AN ORGANIZED METHOD.
A STRAND OF DNA IS COMPRISED OF THOUSANDS OF MOLECULES OF NUCLEIC ACIDS.
SPECIFIC GROUPINGS OF THOSE NUCLEIC ACIDS, WHICH ENCODE A SPECIFIC PHYSICAL CHARACTERISTIC, ARE CALLED GENES.
GENES ARE UNITS OF INFORMATION.
THE STRANDS OF DNA-- CONTAINING ALL THE GENETIC INFORMATION--
--ARE NEATLY PACKED INTO STRUCTURES KEPT WITHIN THE CELL CALLED CHROMOSOMES.
A COMPLETE SET OF AN ORGANISM'S GENES-- THE TOTALITY OF ALL THE INFORMATION NEEDED TO CREATE ANOTHER IDENTICAL ORGANISM --
--IS CALLED A GENOME.
EARTHLY ORGANISMS CAN HAVE ANYWHERE FROM HUNDREDS TO TENS OF THOUSANDS OF GENES WITHIN THEIR GENOME, REGARDLESS OF STRUCTURAL COMPLEXITY!
ROUNDWORM: 20,000 GENES
CHICKEN: 23,000 GENES
MAN: 20-25,000 GENES

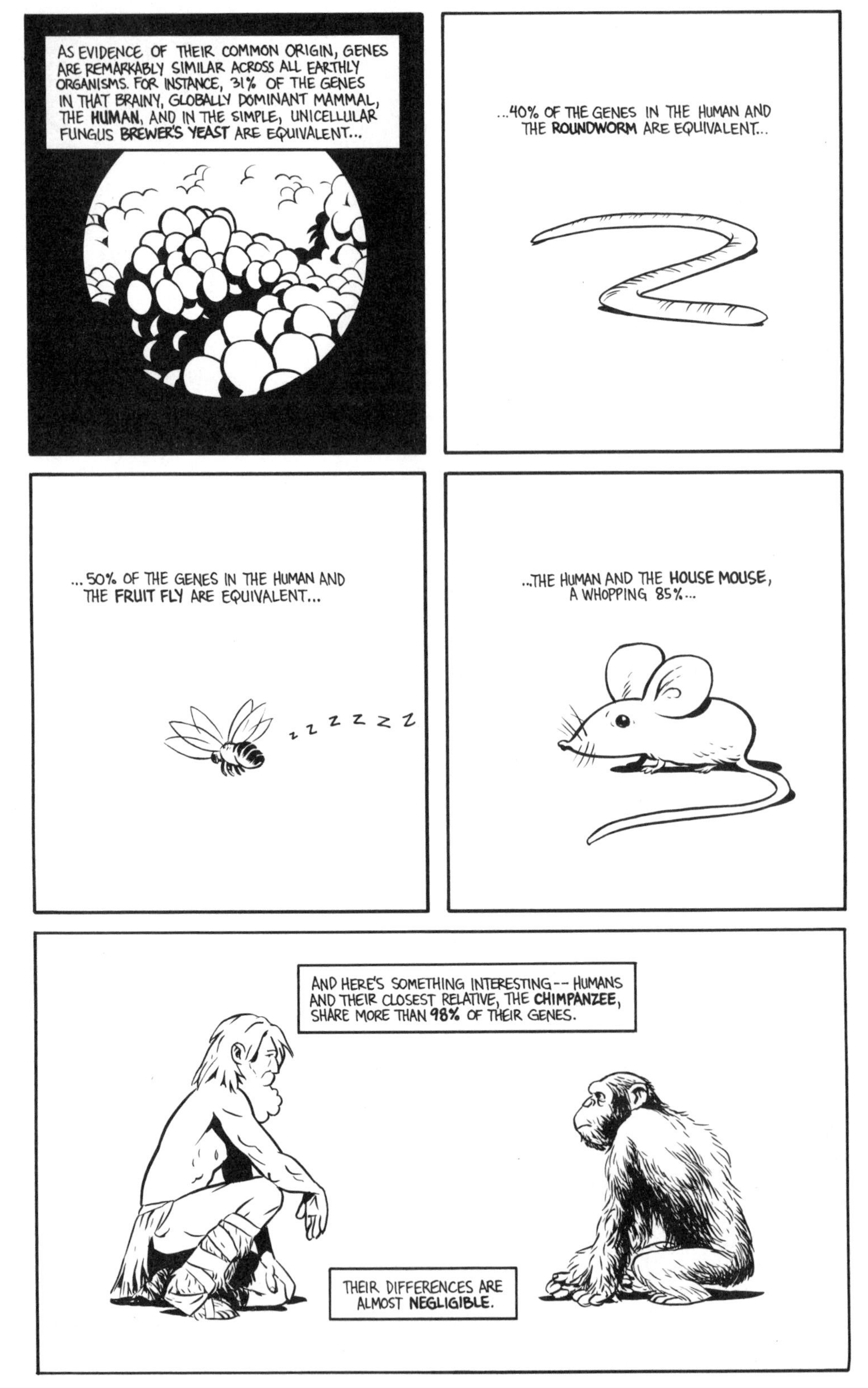
AS EVIDENCE OF THEIR COMMON ORIGIN, GENES ARE REMARKABLY SIMILAR ACROSS ALL EARTHLY ORGANISMS. FOR INSTANCE, 31% OF THE GENES IN THAT BRAINY, GLOBALLY DOMINANT MAMMAL, THE HUMAN, AND IN THE SIMPLE, UNICELLULAR FUNGUS BREWER'S YEAST ARE EQUIVALENT...
...40% OF THE GENES IN THE HUMAN AND THE ROUNDWORM ARE EQUIVALENT...
...50% OF THE GENES IN THE HUMAN AND THE FRUIT FLY ARE EQUIVALENT...
z z z z z z
...THE HUMAN AND THE HOUSE MOUSE, A WHOPPING 85%...
AND HERE'S SOMETHING INTERESTING -- HUMANS AND THEIR CLOSEST RELATIVE, THE CHIMPANZEE, SHARE MORE THAN 98% OF THEIR GENES.
THEIR DIFFERENCES ARE ALMOST NEGLIGIBLE.

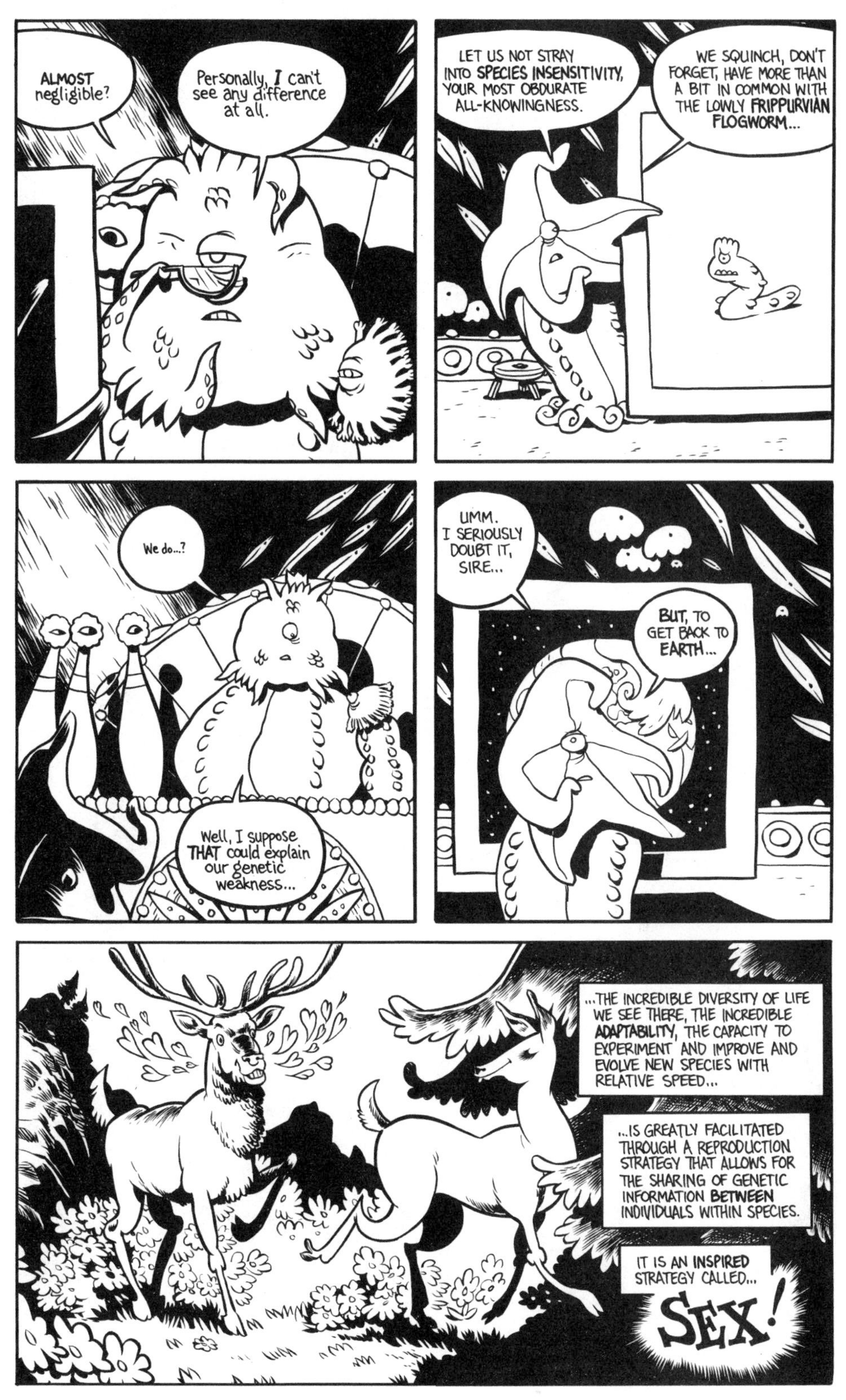
ALMOST negligible?
Personally, I can't see any difference at all.
LET US NOT STRAY INTO SPECIES INSENSITIVITY, YOUR MOST OBDURATE ALL-KNOWINGNESS.
WE SQUINCH, DON'T FORGET, HAVE MORE THAN A BIT IN COMMON WITH THE LOWLY FRIPPURVIAN FLOGWORM...
We do...?
Well, I suppose THAT could explain our genetic weakness...
UMM. I SERIOUSLY DOUBT IT, SIRE...
BUT, TO GET BACK TO EARTH...
...THE INCREDIBLE DIVERSITY OF LIFE WE SEE THERE, THE INCREDIBLE ADAPTABILITY, THE CAPACITY TO EXPERIMENT AND IMPROVE AND EVOLVE NEW SPECIES WITH RELATIVE SPEED...
...IS GREATLY FACILITATED THROUGH A REPRODUCTION STRATEGY THAT ALLOWS FOR THE SHARING OF GENETIC INFORMATION BETWEEN INDIVIDUALS WITHIN SPECIES.
IT IS AN INSPIRED STRATEGY CALLED...
SEX!

SHARING genetic information? Ugh.
That seems like it might be--unpleasant. INTRUSIVE, to say the least.
WELL, THERE ARE SOME DISADVANTAGES. AND LIFE ON EARTH DID BEGIN BY REPRODUCING JUST AS WE DO-- ASEXUALLY, AS IT'S CALLED THERE.
A LOT OF LOWER ORGANISMS THERE STILL PASS THEIR GENETIC HERITAGE TO THE NEXT GENERATION BY ESSENTIALLY DUPLICATING THEIR PERSONAL GENETIC MAKEUP--
--A PROCESS REFERRED TO ON EARTH AS CLONING.
Well, sure!
Like little JUNIOR here...
...Floorsh number 736-- may the disorder not strike him!
OH, NEVER, YOUR MOST PRODUCTIVE FECUNDITY!
...BUT MOST OF THE MORE COMPLEX ORGANISMS ON EARTH HAVE CREATED A DUALITY WITHIN THEIR SPECIES. TWO VARIANTS OF A SPECIES HAVE EVOLVED -- TWO SEXES-- CALLED MALE AND FEMALE.
THE MALE AND THE FEMALE MUST COMBINE THEIR GENETIC MATERIAL TO SUCCESSFULLY REPRODUCE.
THIS GUARANTEES THAT THE OFFSPRING WILL BE A REPLICATE OF NEITHER OF THE TWO PARENTS, BUT A COMPLETELY UNIQUE COMBINATION OF VARIOUS TRAITS INHERITED FROM THE TWO!
EVERY OFFSPRING OF SEXUAL REPRODUCTION IS SOMETHING NEW--AN EXPERIMENT IN INDEPENDENTLY AND RANDOMLY RESHUFFLING THE PARENT GENES!

Seems needlessly complex.
And who would want a child who isn't exactly as expected?
A SPECIES UNDER PRESSURE IN A VERY COMPETITIVE ENVIRONMENT, THAT'S WHO!
IF YOUR SPECIES IS BEING OUT-COMPETED BY ANOTHER SPECIES FOR--SAY--A LIMITED FOOD SOURCE, YOU'D BETTER **HOPE** YOUR OFF-SPRING GROW STRONGER, OR MORE RESOURCEFUL, THAN THE COMPETITION!
SEXUAL REPRODUCTION MARKEDLY SPEEDS UP THE RATE OF CHANGE WITHIN A SPECIES BECAUSE IT OFFERS MANY MORE VARIABLES, IN THE FORM OF GENES DESCENDING FROM **TWO** LINEAGES WITH ALL **THEIR** VARIATIONS.
ULTIMATELY, PROBABILITY DICTATES THAT SOME OF THOSE INFINITELY VARIABLE COMBINATIONS OF PARENTAL TRAITS WILL PRODUCE POSITIVE CHANGES-- **COMPETITIVE ADVANTAGES** FOR THE SPECIES.
OF COURSE, IF THE COMPETITION IS REPRODUCING SEXUALLY, TOO, THEN THEY'RE PROBABLY KEEPING UP WITH YOU ON THE EVOLUTIONARY BATTLEFIELD.
THAT'S THE DOWNSIDE OF SEX-- IT TAKES AN AWFUL LOT OF ENERGY, AND THERE IS A LOT OF WASTE. MOST NEW COMBINATIONS OF GENES DON'T OFFER ANY USEFUL ADVANTAGE, AND SOME COMBINATIONS ARE HURTFUL...
...BUT ON THE VERY COMPETITIVE PLANET EARTH, OVER THE LONG HAUL, IT HAS PROVEN TO BE THE PROGRESSIVE, WINNING STRATEGY.
SPLORT!!
OUCH.
ASEXUAL REPRODUCTION IS SAFE-- YOU KNOW WHAT YOU'RE GOING TO GET-- BUT THAT CONSISTENCY LEADS TO A COMPARATIVELY **SLOW** RATE OF EVOLUTION.
PRIMATES EMPLOY **SEXUAL REPRODUCTION** --AND IN **HUMANS**, THEY HAVE DEVELOPED A WORLD-DOMINATING INTELLIGENCE...
...IN LESS THAN 5 MILLION YEARS!

Less than five...
That is fast, isn't it?
IT TOOK US A LOT LONGER TO GROW OUR BRAINS.
Okay, I think I see where this is going...
You REALLY believe something of this sex business can help us?
YES, I DO, YOUR MOST INCISIVE PERCEPTIVENESS.
I WAS SENT TO EARTH WITH A SPECIFIC MISSION-- TO LEARN HOW THAT PLANET'S LIFE DEALS WITH INHERITED DISEASE...
I am WELL aware of the mission, Bloort 183! The genetic disorder afflicting my household is something I must deal with EVERY DAY.
The rest of Squinchdom, too-- I feel for ALL my subjects.
And I don't like the word DISEASE.
Y-YES, SIRE...
THE FACT IS THAT WHILE OUR CONSERVATIVE METHOD OF SELF-REPRODUCING HAS KEPT OUR HEALTH REMARKABLY STABLE FOR MOST OF OUR EXISTENCE, IT HAS ALSO INHIBITED OUR ABILITY TO RESPOND TO NEW CHALLENGES...
...AS WE NOW SEE WITH THE GROWTH OF THESE AWFUL GENETIC ...DISORDERS, AGAINST WHICH WE HAVE NO DEFENSE.
WE HAD HOPED THAT MY STUDY OF EARTH LIFE MIGHT OFFER A SOLUTION-- AND I THINK IT HAS!
I'M CONVINCED THAT THE ADVANTAGES THAT COME WITH EXCHANGING GENES WILL BE USEFUL IN STRENGTHENING SQUINCH BIOLOGY!
IT IS NOT NECESSARY TO HAVE TWO DIFFERING SETS OF REPRODUCTIVE ORGANS. SEXUAL REPRODUCTION CAN BE ACHIEVED ARTIFICIALLY, AS YOU WILL SEE.

THE HUMAN ORGANISMS HAVE LEARNED AMAZING THINGS ABOUT THEIR GENETIC LEGACY AND ARE ON THE VERGE OF CREATING A REVOLUTION IN GENETIC HEALTH FOR THEMSELVES.
THERE IS MUCH WE CAN BORROW FROM THEM.
THEY CLEARLY HAVE HIGH EXPECTATIONS FOR THEMSELVES...
IN THEIR SCIENTIFIC LANGUAGE, HUMANS DESIGNATE THEMSELVES AS HOMO SAPIENS, WHICH MEANS "WISE, KNOWING MAN."
ALTHOUGH, IN A SIMILAR PARLANCE, I WOULD CHOOSE TO CALL THEM SOPHOMORES --
"WISE FOOLS," ONLY PARTWAY THROUGH THEIR SELF-EDUCATION, IN CONSTANT DANGER OF MAKING TERRIBLE MISTAKES.
THEY WRESTLE CONTINUALLY WITH THE ETHICAL AND PRACTICAL DILEMMAS INHERENT IN THEIR EVERY DISCOVERY AND INVENTION...
...WHICH FOLLOW ONE UPON THE OTHER WITH TREMENDOUS SPEED.
Because of -- SEX?
NOT EXACTLY, SIRE...
ACTUALLY-- YES, YOUR GIGANTIC LORDLINESS!
SEX DOES HAVE A LOT TO DO WITH IT!
IT WAS SEXUAL REPRODUCTION THAT FIRST LED MAN TO BEGIN HIS UNDERSTANDING OF THE POSSIBILITIES OF GENETICS...
SEVERAL THOUSAND YEARS AGO, A HUMAN -- MAYBE A KEEPER OF DOMESTICATED ORGANISMS -- FIRST RECOGNIZED A BASIC, BUT IMPORTANT, PRINCIPLE: LIKE BEGETS LIKE.
SHEEP ALWAYS PRODUCE SHEEP, AND GRAPES, MORE GRAPES. GRAPES NEVER PRODUCE SHEEP.

THE HUMANS' GRASP OF THE TRUISM THAT **ORGANISMS REPRODUCE THEMSELVES** LED THEM TO OBSERVE THAT SPECIFIC TRAITS COULD BE PASSED FROM PARENT TO OFFSPRING.
THIS LED TO THE **SELECTIVE BREEDING** OF ANIMALS AND PLANTS WITH CHARACTERISTICS THEY FOUND DESIRABLE...
...AND THE DELIBERATE EXCLUSION OF THOSE WITH UNDESIRABLE TRAITS.
OBSERVATION SHOWED WHERE HUMANS WERE RIGHT AND WHERE THEY WERE WRONG.
NATURE CHANGES SPECIES BLINDLY, OVER LONG PERIODS OF TIME, BY PURE TRIAL AND ERROR. HUMANS CONSCIOUSLY MANIPULATED GENES WITH A CRITICAL, PURPOSE-DRIVEN EYE AND SAW RESULTS WITH REWARDING SPEED.
THIS WAS THE **BEGINNING** OF THE UNDERSTANDING OF **GENETICS**.
BUT THAT WAS PRETTY MUCH **IT** FOR THOUSANDS OF YEARS. ONGOING ATTEMPTS TO GRASP THE **UNDERLYING MECHANICS** OF REPRODUCTION SPROUTED SOME INTERESTING THEORIES-- **SPONTANEOUS GENERATION** WAS POPULARLY ACCEPTED FOR A LONG TIME--
Behold! Dirty sheets give rise to bedbugs!
--BUT THEY WERE ALMOST ALWAYS STEPS **BACKWARD**.
IT WAS LESS THAN 200 YEARS AGO THAT TWO PARTICULAR HOMO SAPIENS--HUMANS, THAT IS--FINALLY STARTED THINGS ON THE RIGHT TRACK.
CHARLES DARWIN and GREGOR MENDEL
ACCURATELY CONCLUDED THAT SPECIES **EVOLVE** BASED ON THE ABILITIES OF INDIVIDUAL ORGANISMS TO **SURVIVE** AND **REPRODUCE**.
ACCURATELY OBSERVED AND DESCRIBED **HOW** TRAITS ARE PASSED FROM ONE GENERATION TO THE NEXT.
USING THE **SCIENTIFIC METHOD** AND **CONTROLLED EXPERIMENTS**, MENDEL OBSERVED GENERATIONS OF CAREFULLY CULTIVATED PEA PLANTS AND FROM THEM DETERMINED THE **LAWS OF INHERITANCE**.
HE FOUNDED THE **SCIENCE** OF GENETICS, ALTHOUGH IT WOULD BE DECADES BEFORE HIS BREAKTHROUGHS WERE PROPERLY RECOGNIZED.

BUT ONCE SCIENCE DID ACCEPT MENDEL'S GENES, THE MYSTERIES OF THE MECHANICS BEHIND HEREDITY WERE SOLVED WITH REMARKABLE RAPIDITY.
JAMES WATSON
FRANCIS CRICK
THE RELATIONSHIP BETWEEN GENES AND CHROMOSOMES WAS DISCOVERED -- DNA WAS DESCRIBED -- AND THE COMPOSITION AND STRUCTURE OF DNA WAS UNRAVELED...
... AND NOW THE GENETIC LANGUAGE ITSELF IS BEING TRANSLATED AND INTERPRETED!
HUMANITY IS LEARNING TO READ THE BOOK OF LIFE!
PERHAPS THE CROWNING ACHIEVEMENT HAS BEEN THE HUMAN GENOME PROJECT --
BIOLOGY
PHYSICS
ETHICS
INFORMATICS
ENGINEERING
CHEMISTRY
-- A MASSIVE EFFORT THAT HAS MAPPED THE ENTIRETY OF HUMAN DNA -- ALL 3 BILLION BASE PAIRS (I'LL EXPLAIN LATER).
All right -- so they've gotten to know themselves intimately.
How does this help us deal with our problems?
THE KNOWLEDGE THEY'VE GAINED ISN'T JUST FOR ITS OWN SAKE.
THEY'RE APPLYING WHAT THEY'VE LEARNED.
THEY'RE CHANGING THEIR WORLD -- AND THEMSELVES -- THROUGH THEIR GROWING UNDER-STANDING OF GENETICS.
MAYBE MOST IMPORTANT, THEY ARE LEARNING HOW TO ALLEVIATE THE SUFFERING THAT COMES WHEN DEFECTIVE OR BADLY MATCHED GENES CAUSE DISORDERS AND DISEASES, OR WHEN CHROMOSOMES ARE DAMAGED.

BASIC KNOWLEDGE OF HEREDITY HAS TAUGHT THEM TO TRACE THEIR ANCESTRIES. AN UNDERSTANDING OF THE SPECIFIC GENETIC TRAITS IN A FAMILY HISTORY CAN HELP INDIVIDUALS ESTIMATE THEIR CHANCES OF PASSING ON THOSE TRAITS TO THEIR CHILDREN -- QUITE IMPORTANT WHEN IT COMES TO **AVOIDING** HEREDITARY GENETIC DISORDERS.
IT MIGHT BE HELPFUL IF **WE** COULD LEARN TO CALCULATE THE DANGERS INHERENT IN **OUR** LINEAGES!
SOME GENETIC PROBLEMS POP UP **WITHOUT** A HISTORY. GENETIC CHANGE -- **MUTATION** -- OCCURS WITHIN INDIVIDUALS, AND CAN BE EITHER GOOD OR BAD.
CANCERS ARE DISEASES OF DNA THAT HAS BEEN DAMAGED BY MUTATION.
NORMAL
CANCEROUS
HUMANITY IS USING ITS KNOWLEDGE OF DNA TO CONSTRUCT THERAPIES FOR MANY OF THESE GENETIC PROBLEMS.
ONE OF THE MOST PROMISING THERAPIES INVOLVES CONVERTING **VIRUSES** TO DELIVERY **VECTORS** CARRYING HEALTHY GENES TO AFFLICTED CELLS.
VIRUSES NATURALLY **HIJACK** CELLS TO REPRODUCE THEMSELVES.
HEALTHY GENES
HUMAN GENETICISTS HAVE LEARNED TO TAME THESE TINY CAPSULES, LOAD THEM WITH HEALTHY GENES...
...AND LAUNCH THEM AT THE CELLS OF THE APPROPRIATE TISSUE. THE VIRUSES, DOING WHAT COMES NATURALLY, DELIVER THEIR PAYLOAD.
Le' me at 'im!
ZOOM
BUT INSTEAD OF REPRODUCING THEMSELVES, THEY INTRODUCE NEW, HEALTHY HUMAN GENES TO REPLACE THE DAMAGED ONES. **AMAZING!** THIS IS CALLED **RECOMBINANT DNA TECHNOLOGY.**
THIS TECHNOLOGY IS USED FOR OTHER PURPOSES AS WELL. HUMAN DNA IS BEING RECOMBINED WITH **BACTERIA** -- WHICH REPRODUCE QUICKLY AND UNDER HIGHLY CONTROLLED CONDITIONS -- TO PRODUCE LARGE AND AFFORDABLE QUANTITIES OF PREVIOUSLY DIFFICULT-TO-MASS-PRODUCE DRUGS SUCH AS **INSULIN** AND **INTERFERON**, WHICH ARE USED TO COUNTER THE EFFECTS OF DAMAGED GENES.
eep!
E. COLI
E. COLI
E. COLI
E. COLI
THE TECHNIQUE OF RECOMBINANT DNA HAS EVEN BEEN HARNESSED TO CREATE CERTAIN **PLANTS** THAT PRODUCE HUMAN **ANTIBODIES!** THESE ARE EASILY HARVESTED AND USED, ONCE AGAIN, TO FIGHT DISEASES.
?!
WHAT NEXT?

HUMANITY HAS LONG UNDERSTOOD THE PRINCIPLE OF CLONING, AND HAS EMPLOYED IT FOR CENTURIES AS A TOOL IN KEEPING CONSISTENT QUALITIES IN AGRICULTURAL PRODUCE, SUCH AS THE FRUIT CALLED APPLES.
BUT IT WASN'T UNTIL RECENTLY THAT THEIR SKILLS ALLOWED THEM TO BEGIN CLONING ANIMALS. ALTHOUGH THEY'VE HAD LIMITED SUCCESS UP UNTIL NOW, IT IS PROBABLE THAT THEY WILL SOONER OR LATER PERFECT THIS TECHNIQUE.
AND LET ME TELL YOU, THERE ARE SOME HEATED DISCUSSIONS ABOUT THIS BREWING ON EARTH!
SOME HUMANS HAIL CLONING AS AN ANSWER TO THEIR LIMITED LIFE SPANS-- A POSSIBILITY FOR INDIVIDUALS TO KEEP BANKS OF THEIR OWN CLONED ORGANS AS REPLACEMENTS FOR THE WORN-OUT ORIGINALS.
BOB
BOB'S XTRA HEART
BOB'S XTRA KIDNEY
BOB'S XTRA LIVER
BOB'S XTRA TESTES
BOB'S XTRA LUNG
BOB'S XTRA SPLEEN
SOME EVEN BELIEVE THAT DNA ITSELF MAY BE CLONED TO INDEFINITELY DELAY THEIR AGING PROCESS!
AS YOU MIGHT IMAGINE, THIS IS RAISING PLENTY OF ETHICAL QUESTIONS--AS DO CERTAIN HUMANS' DREAMS OF PURPOSE-FULLY CHOOSING TO GENETICALLY ENGINEER DESIRED TRAITS INTO FUTURE GENERATIONS.
...BABIES EXPRESSLY DESIGNED TO ORDER...
...BELOVED PETS CLONED TO LAST FOREVER.
BUT THE HUMANS' EFFORTS TO MASTER DNA ARE CLEARING UP DILEMMAS AS WELL.
CRIMES ARE BEING SOLVED AND LEGAL ISSUES UNTANGLED WITH THE USE OF DNA PROFILING, WHERE PHYSICAL EVI-DENCE IS ANALYZED FOR THE GENETIC "FINGERPRINTS" OF INDIVIDUALS.
FINALLY, THE GROWING UNDER-STANDING OF THEIR GENOME HAS UNLOCKED MANY OF THE MYSTERIES OF THEIR PREHISTORIC PAST, ALLOWING THEM A GLIMPSE INTO THEIR ORIGINS ON THE CONTINENT CALLED AFRICA AS WELL AS THEIR SUCCESSFUL DISPERSION AROUND THE GLOBE.
AS THEY LEARN ABOUT THE GENETIC STRENGTHS AND WEAKNESSES OF THEIR GEOGRAPHICALLY DIVERSE POPULATIONS...
ARABIA

...THEY GAIN A DEEPER UNDERSTANDING OF THEMSELVES -- THEIR ABILITIES, THEIR LIMITS, THEIR POTENTIAL.
THEY ARE A SPECIES CONSTANTLY PUSHING TO IMPROVE THEMSELVES AND THEIR WORLD...
...AND THEY STRUGGLE EVERY **INCH** OF THE WAY WITH ALL THE COMPLEXITIES THAT ENTAILS.
THE RELATIVE VALUES OF SHORT-RANGE GAINS IN HEALTH WEIGHED AGAINST LONG-RANGE GENETIC DANGERS--
-- IMMEDIATE PROFIT WEIGHED AGAINST UNIMAGINABLE BACKLASHES IN THE DISTANT FUTURE...
...WHEW!
LIFE ON EARTH AND THE DNA THAT MAKES IT WORK ARE REMARKABLE THINGS.
HUMANITY IS A REMARKABLE SPECIES, AND I REPEAT-- WE CAN LEARN FROM THEM. ENOUGH, I THINK, TO HELP OURSELVES.
NOW THAT I'VE CONCLUDED MY **INTRODUCTION**, YOUR MOST DISCRIMINATING LORDLINESS, DO I HAVE YOUR PERMISSION TO CONTINUE?
Of **COURSE**! I've heard enough to be intrigued.
Get **ON** with it, for pity's sake...
We have much to learn, and little time.
VERY GOOD, THEN, SIRE.
THERE ARE MANY PLACES WE COULD START, BUT I SUGGEST WE BEGIN WITH THE **BASICS**, WITH THE ELEMENTS THAT COMPRISE US **ALL**...
...NO MATTER WHAT **WORLD** WE COME FROM.

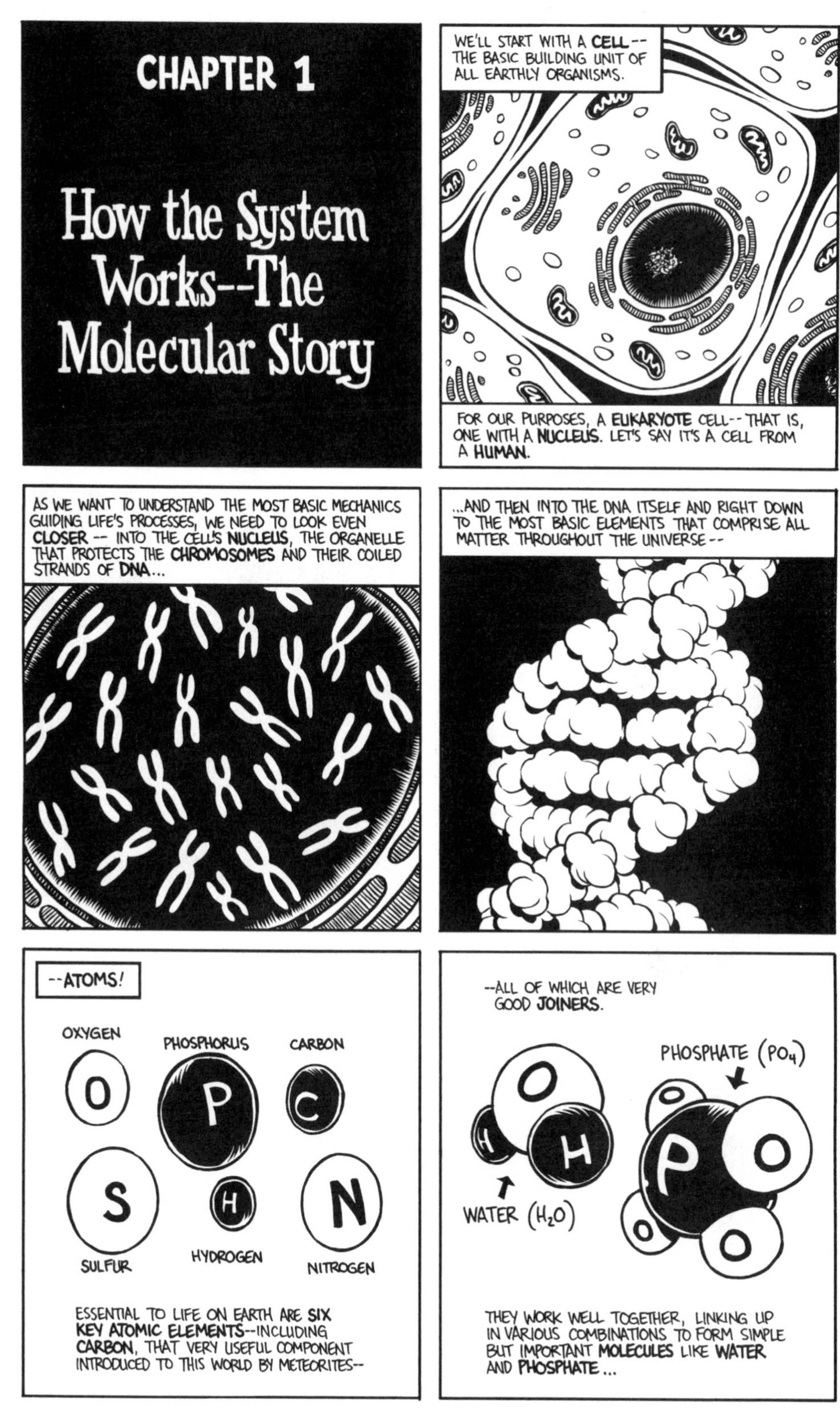
CHAPTER 1
How the System Works--The Molecular Story
WE'LL START WITH A CELL -- THE BASIC BUILDING UNIT OF ALL EARTHLY ORGANISMS.
FOR OUR PURPOSES, A EUKARYOTE CELL -- THAT IS, ONE WITH A NUCLEUS. LET'S SAY IT'S A CELL FROM A HUMAN.
AS WE WANT TO UNDERSTAND THE MOST BASIC MECHANICS GUIDING LIFE'S PROCESSES, WE NEED TO LOOK EVEN CLOSER -- INTO THE CELL'S NUCLEUS, THE ORGANELLE THAT PROTECTS THE CHROMOSOMES AND THEIR COILED STRANDS OF DNA...
...AND THEN INTO THE DNA ITSELF AND RIGHT DOWN TO THE MOST BASIC ELEMENTS THAT COMPRISE ALL MATTER THROUGHOUT THE UNIVERSE --
--ATOMS!
OXYGEN
PHOSPHORUS
CARBON
O
P
C
S
H
N
SULFUR
HYDROGEN
NITROGEN
ESSENTIAL TO LIFE ON EARTH ARE SIX KEY ATOMIC ELEMENTS--INCLUDING CARBON, THAT VERY USEFUL COMPONENT INTRODUCED TO THIS WORLD BY METEORITES--
--ALL OF WHICH ARE VERY GOOD JOINERS.
PHOSPHATE (PO_4)
WATER (H_2O)
THEY WORK WELL TOGETHER, LINKING UP IN VARIOUS COMBINATIONS TO FORM SIMPLE BUT IMPORTANT MOLECULES LIKE WATER AND PHOSPHATE...

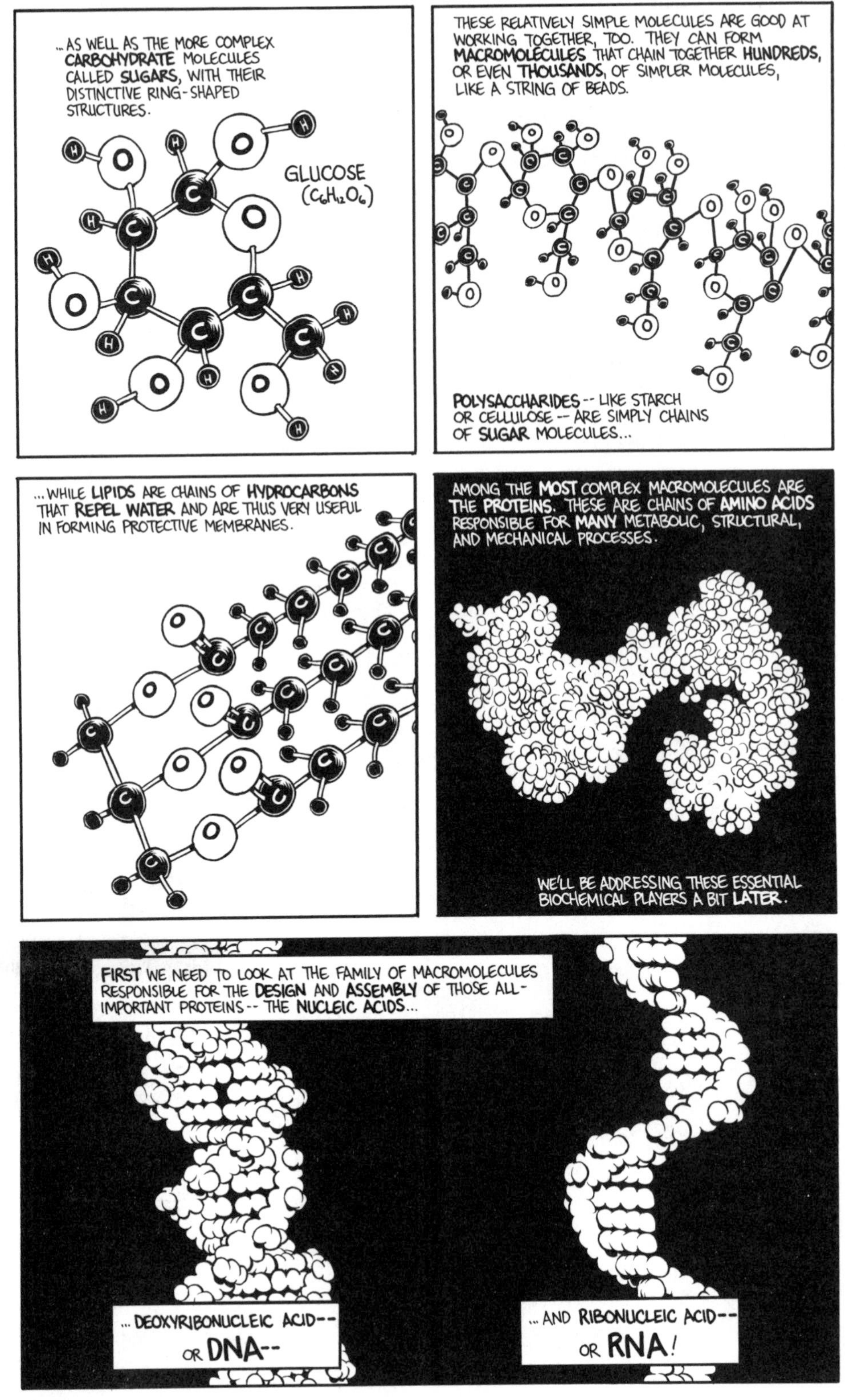
...AS WELL AS THE MORE COMPLEX CARBOHYDRATE MOLECULES CALLED SUGARS, WITH THEIR DISTINCTIVE RING-SHAPED STRUCTURES.
GLUCOSE ($C_6H_{12}O_6$)
THESE RELATIVELY SIMPLE MOLECULES ARE GOOD AT WORKING TOGETHER, TOO. THEY CAN FORM MACROMOLECULES THAT CHAIN TOGETHER HUNDREDS, OR EVEN THOUSANDS, OF SIMPLER MOLECULES, LIKE A STRING OF BEADS.
POLYSACCHARIDES -- LIKE STARCH OR CELLULOSE -- ARE SIMPLY CHAINS OF SUGAR MOLECULES...
...WHILE LIPIDS ARE CHAINS OF HYDROCARBONS THAT REPEL WATER AND ARE THUS VERY USEFUL IN FORMING PROTECTIVE MEMBRANES.
AMONG THE MOST COMPLEX MACROMOLECULES ARE THE PROTEINS. THESE ARE CHAINS OF AMINO ACIDS RESPONSIBLE FOR MANY METABOLIC, STRUCTURAL, AND MECHANICAL PROCESSES.
WE'LL BE ADDRESSING THESE ESSENTIAL BIOCHEMICAL PLAYERS A BIT LATER.
FIRST WE NEED TO LOOK AT THE FAMILY OF MACROMOLECULES RESPONSIBLE FOR THE DESIGN AND ASSEMBLY OF THOSE ALL-IMPORTANT PROTEINS -- THE NUCLEIC ACIDS...
...DEOXYRIBONUCLEIC ACID-- OR DNA--
...AND RIBONUCLEIC ACID-- OR RNA!

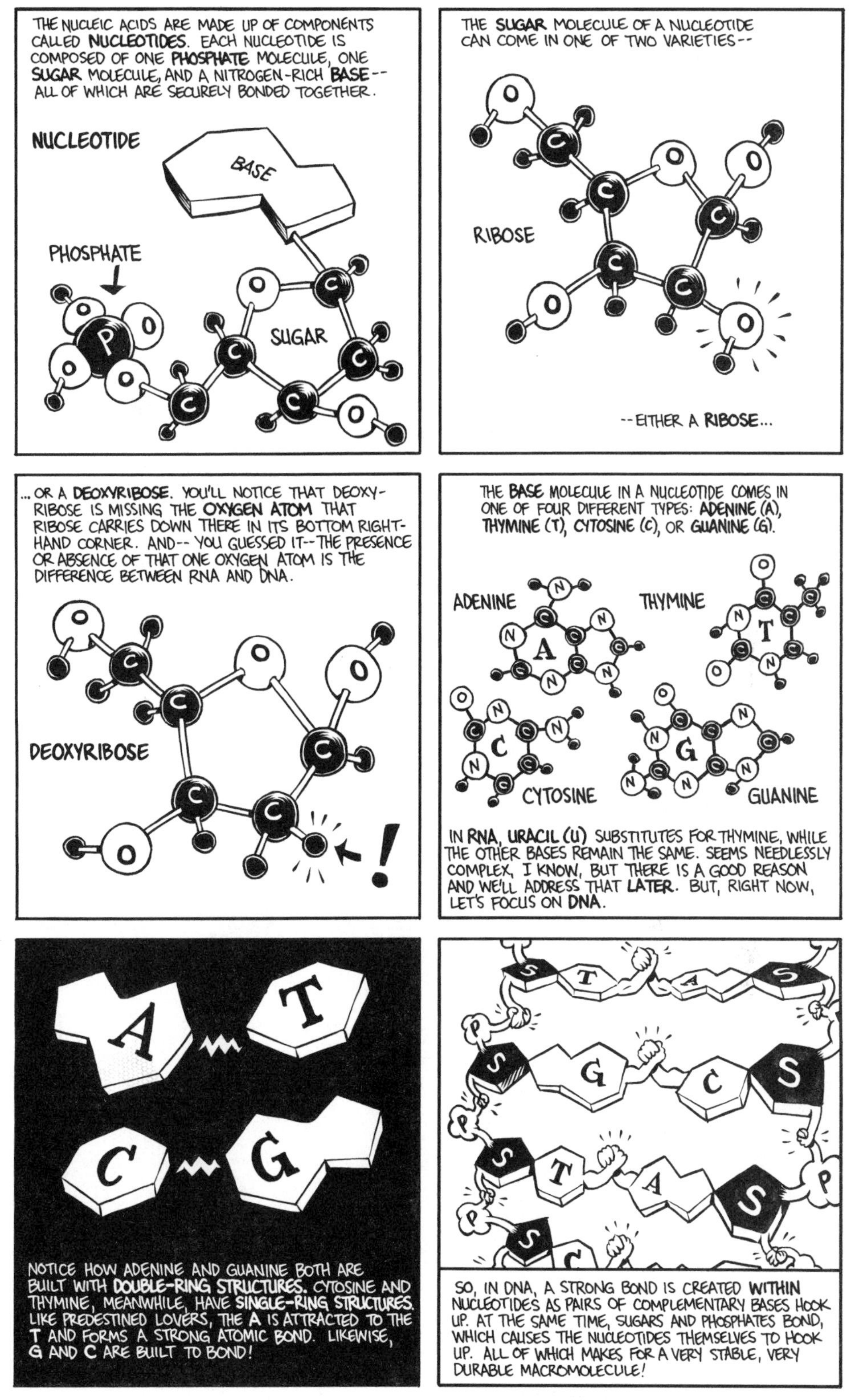

THE NUCLEIC ACIDS ARE MADE UP OF COMPONENTS CALLED **NUCLEOTIDES**. EACH NUCLEOTIDE IS COMPOSED OF ONE **PHOSPHATE** MOLECULE, ONE **SUGAR** MOLECULE, AND A NITROGEN-RICH **BASE**-- ALL OF WHICH ARE SECURELY BONDED TOGETHER.
NUCLEOTIDE
BASE
PHOSPHATE
SUGAR
THE **SUGAR** MOLECULE OF A NUCLEOTIDE CAN COME IN ONE OF TWO VARIETIES--
RIBOSE
--EITHER A **RIBOSE**...
...OR A **DEOXYRIBOSE**. YOU'LL NOTICE THAT DEOXYRIBOSE IS MISSING THE **OXYGEN ATOM** THAT RIBOSE CARRIES DOWN THERE IN ITS BOTTOM RIGHT-HAND CORNER. AND-- YOU GUESSED IT-- THE PRESENCE OR ABSENCE OF THAT ONE OXYGEN ATOM IS THE DIFFERENCE BETWEEN RNA AND DNA.
DEOXYRIBOSE
!
THE **BASE** MOLECULE IN A NUCLEOTIDE COMES IN ONE OF FOUR DIFFERENT TYPES: **ADENINE** (A), **THYMINE** (T), **CYTOSINE** (C), OR **GUANINE** (G).
ADENINE
THYMINE
CYTOSINE
GUANINE
IN **RNA**, **URACIL** (U) SUBSTITUTES FOR THYMINE, WHILE THE OTHER BASES REMAIN THE SAME. SEEMS NEEDLESSLY COMPLEX, I KNOW, BUT THERE IS A GOOD REASON AND WE'LL ADDRESS THAT **LATER**. BUT, RIGHT NOW, LET'S FOCUS ON **DNA**.
NOTICE HOW ADENINE AND GUANINE BOTH ARE BUILT WITH **DOUBLE-RING STRUCTURES.** CYTOSINE AND THYMINE, MEANWHILE, HAVE **SINGLE-RING STRUCTURES.** LIKE PREDESTINED LOVERS, THE **A** IS ATTRACTED TO THE **T** AND FORMS A STRONG ATOMIC BOND. LIKEWISE, **G** AND **C** ARE BUILT TO BOND!
SO, IN DNA, A STRONG BOND IS CREATED **WITHIN** NUCLEOTIDES AS PAIRS OF COMPLEMENTARY BASES HOOK UP. AT THE SAME TIME, SUGARS AND PHOSPHATES BOND, WHICH CAUSES THE NUCLEOTIDES THEMSELVES TO HOOK UP. ALL OF WHICH MAKES FOR A VERY STABLE, VERY DURABLE MACROMOLECULE!

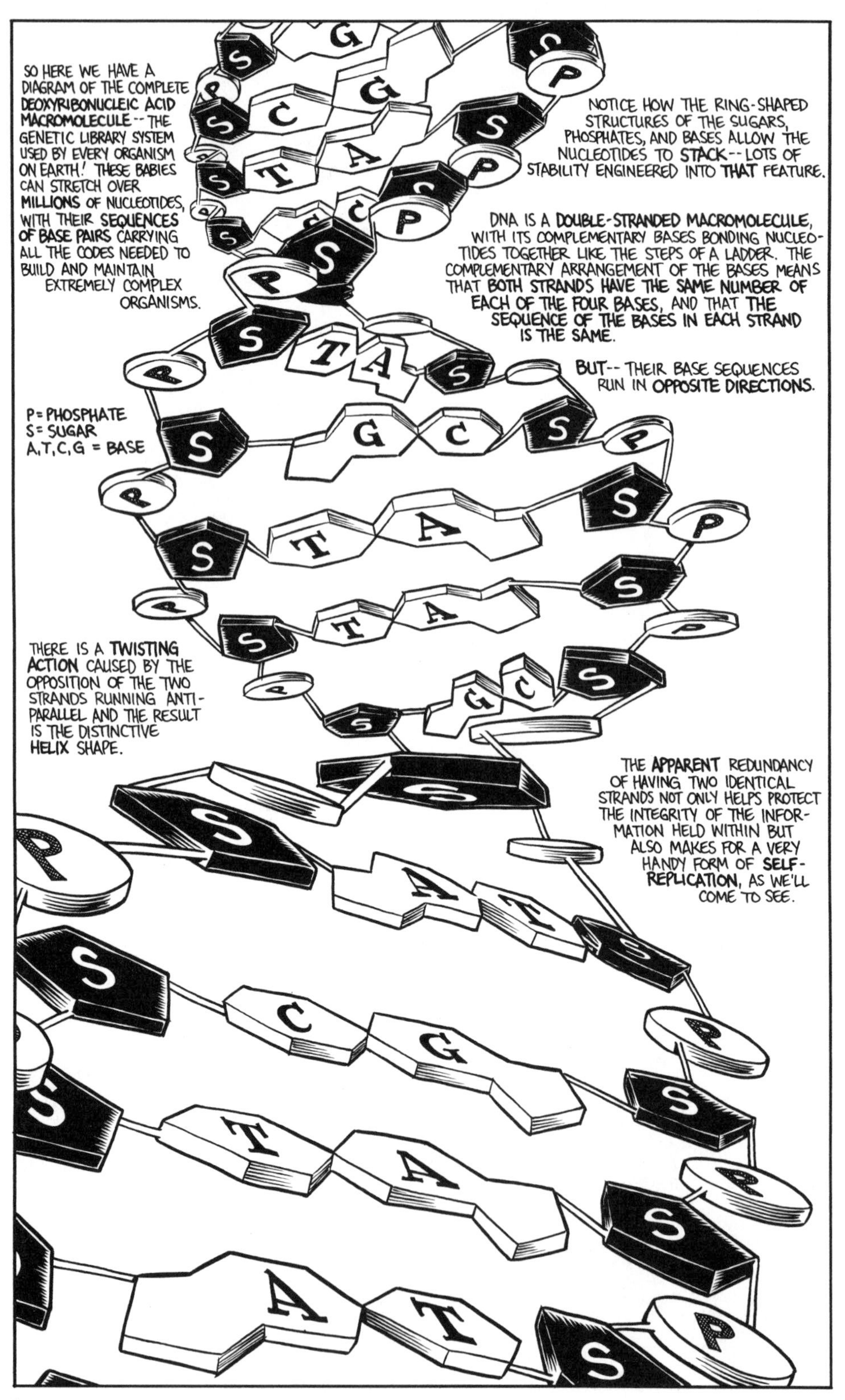
SO HERE WE HAVE A DIAGRAM OF THE COMPLETE DEOXYRIBONUCLEIC ACID MACROMOLECULE -- THE GENETIC LIBRARY SYSTEM USED BY EVERY ORGANISM ON EARTH! THESE BABIES CAN STRETCH OVER MILLIONS OF NUCLEOTIDES, WITH THEIR SEQUENCES OF BASE PAIRS CARRYING ALL THE CODES NEEDED TO BUILD AND MAINTAIN EXTREMELY COMPLEX ORGANISMS.
NOTICE HOW THE RING-SHAPED STRUCTURES OF THE SUGARS, PHOSPHATES, AND BASES ALLOW THE NUCLEOTIDES TO STACK -- LOTS OF STABILITY ENGINEERED INTO THAT FEATURE.
DNA IS A DOUBLE-STRANDED MACROMOLECULE, WITH ITS COMPLEMENTARY BASES BONDING NUCLEOTIDES TOGETHER LIKE THE STEPS OF A LADDER. THE COMPLEMENTARY ARRANGEMENT OF THE BASES MEANS THAT BOTH STRANDS HAVE THE SAME NUMBER OF EACH OF THE FOUR BASES, AND THAT THE SEQUENCE OF THE BASES IN EACH STRAND IS THE SAME.
BUT-- THEIR BASE SEQUENCES RUN IN OPPOSITE DIRECTIONS.
P = PHOSPHATE
S = SUGAR
A, T, C, G = BASE
THERE IS A TWISTING ACTION CAUSED BY THE OPPOSITION OF THE TWO STRANDS RUNNING ANTI-PARALLEL AND THE RESULT IS THE DISTINCTIVE HELIX SHAPE.
THE APPARENT REDUNDANCY OF HAVING TWO IDENTICAL STRANDS NOT ONLY HELPS PROTECT THE INTEGRITY OF THE INFORMATION HELD WITHIN BUT ALSO MAKES FOR A VERY HANDY FORM OF SELF-REPLICATION, AS WE'LL COME TO SEE.

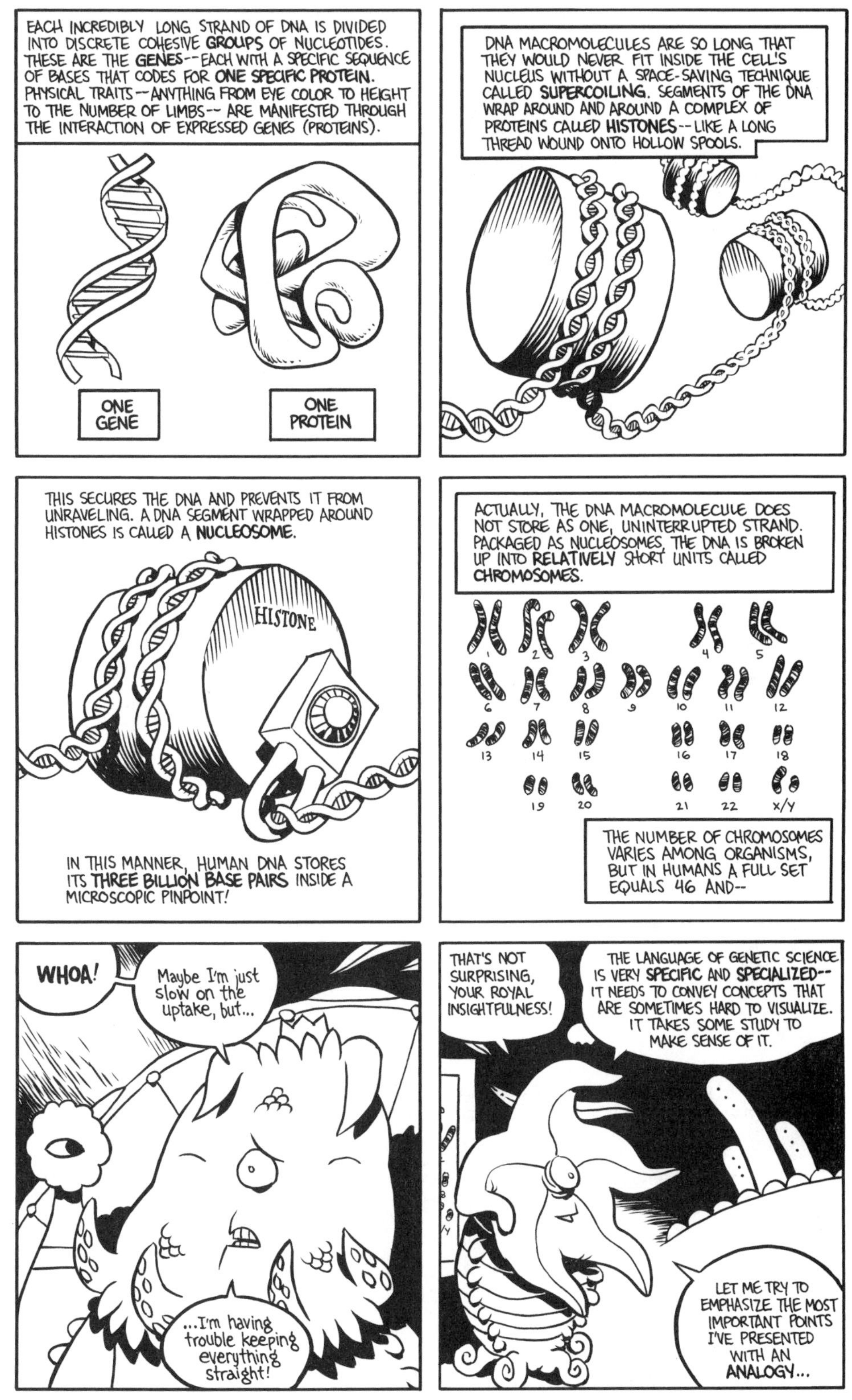
EACH INCREDIBLY LONG STRAND OF DNA IS DIVIDED INTO DISCRETE COHESIVE GROUPS OF NUCLEOTIDES. THESE ARE THE GENES--EACH WITH A SPECIFIC SEQUENCE OF BASES THAT CODES FOR ONE SPECIFIC PROTEIN. PHYSICAL TRAITS--ANYTHING FROM EYE COLOR TO HEIGHT TO THE NUMBER OF LIMBS--ARE MANIFESTED THROUGH THE INTERACTION OF EXPRESSED GENES (PROTEINS).
ONE GENE
ONE PROTEIN
DNA MACROMOLECULES ARE SO LONG THAT THEY WOULD NEVER FIT INSIDE THE CELL'S NUCLEUS WITHOUT A SPACE-SAVING TECHNIQUE CALLED SUPERCOILING. SEGMENTS OF THE DNA WRAP AROUND AND AROUND A COMPLEX OF PROTEINS CALLED HISTONES--LIKE A LONG THREAD WOUND ONTO HOLLOW SPOOLS.
THIS SECURES THE DNA AND PREVENTS IT FROM UNRAVELING. A DNA SEGMENT WRAPPED AROUND HISTONES IS CALLED A NUCLEOSOME.
HISTONE
IN THIS MANNER, HUMAN DNA STORES ITS THREE BILLION BASE PAIRS INSIDE A MICROSCOPIC PINPOINT!
ACTUALLY, THE DNA MACROMOLECULE DOES NOT STORE AS ONE, UNINTERRUPTED STRAND. PACKAGED AS NUCLEOSOMES, THE DNA IS BROKEN UP INTO RELATIVELY SHORT UNITS CALLED CHROMOSOMES.
1 2 3 4 5
6 7 8 9 10 11 12
13 14 15 16 17 18
19 20 21 22 X/Y
THE NUMBER OF CHROMOSOMES VARIES AMONG ORGANISMS, BUT IN HUMANS A FULL SET EQUALS 46 AND--
WHOA!
Maybe I'm just slow on the uptake, but...
...I'm having trouble keeping everything straight!
THAT'S NOT SURPRISING, YOUR ROYAL INSIGHTFULNESS!
THE LANGUAGE OF GENETIC SCIENCE IS VERY SPECIFIC AND SPECIALIZED--IT NEEDS TO CONVEY CONCEPTS THAT ARE SOMETIMES HARD TO VISUALIZE. IT TAKES SOME STUDY TO MAKE SENSE OF IT.
LET ME TRY TO EMPHASIZE THE MOST IMPORTANT POINTS I'VE PRESENTED WITH AN ANALOGY...

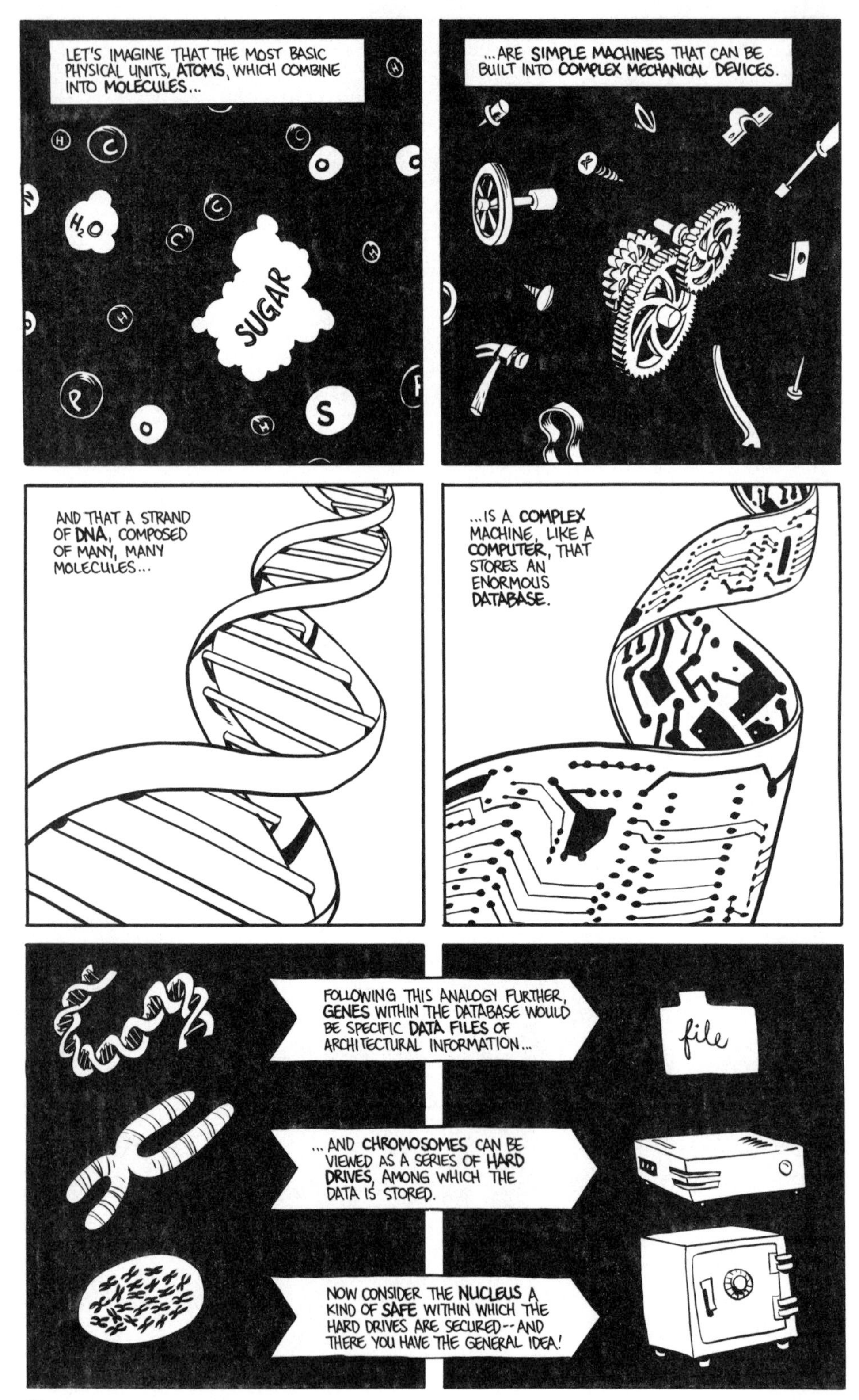
LET'S IMAGINE THAT THE MOST BASIC PHYSICAL UNITS, ATOMS, WHICH COMBINE INTO MOLECULES...
SUGAR
...ARE SIMPLE MACHINES THAT CAN BE BUILT INTO COMPLEX MECHANICAL DEVICES.
AND THAT A STRAND OF DNA, COMPOSED OF MANY, MANY MOLECULES...
...IS A COMPLEX MACHINE, LIKE A COMPUTER, THAT STORES AN ENORMOUS DATABASE.
FOLLOWING THIS ANALOGY FURTHER, GENES WITHIN THE DATABASE WOULD BE SPECIFIC DATA FILES OF ARCHITECTURAL INFORMATION...
file
...AND CHROMOSOMES CAN BE VIEWED AS A SERIES OF HARD DRIVES, AMONG WHICH THE DATA IS STORED.
NOW CONSIDER THE NUCLEUS A KIND OF SAFE WITHIN WHICH THE HARD DRIVES ARE SECURED--AND THERE YOU HAVE THE GENERAL IDEA!

NUCLEUS
Within each cell is a NUCLEUS.
All right -- I can visualize that.
Within each NUCLEUS are discrete structures called CHROMOSOMES.
CHROMOSOMES are composed of DNA MACROMOLECULES.
The DNA, in turn, is divided into GENES -- each of which helps determine a specific physical trait.
A
T
C
G
The GENES themselves are composed of NUCLEOTIDES.
And NUCLEOTIDES are made up of one phosphate molecule, one sugar molecule, and a nitrogen base.
BASE
PHOS-PHATE
SUGAR
YOU'VE GOT IT!

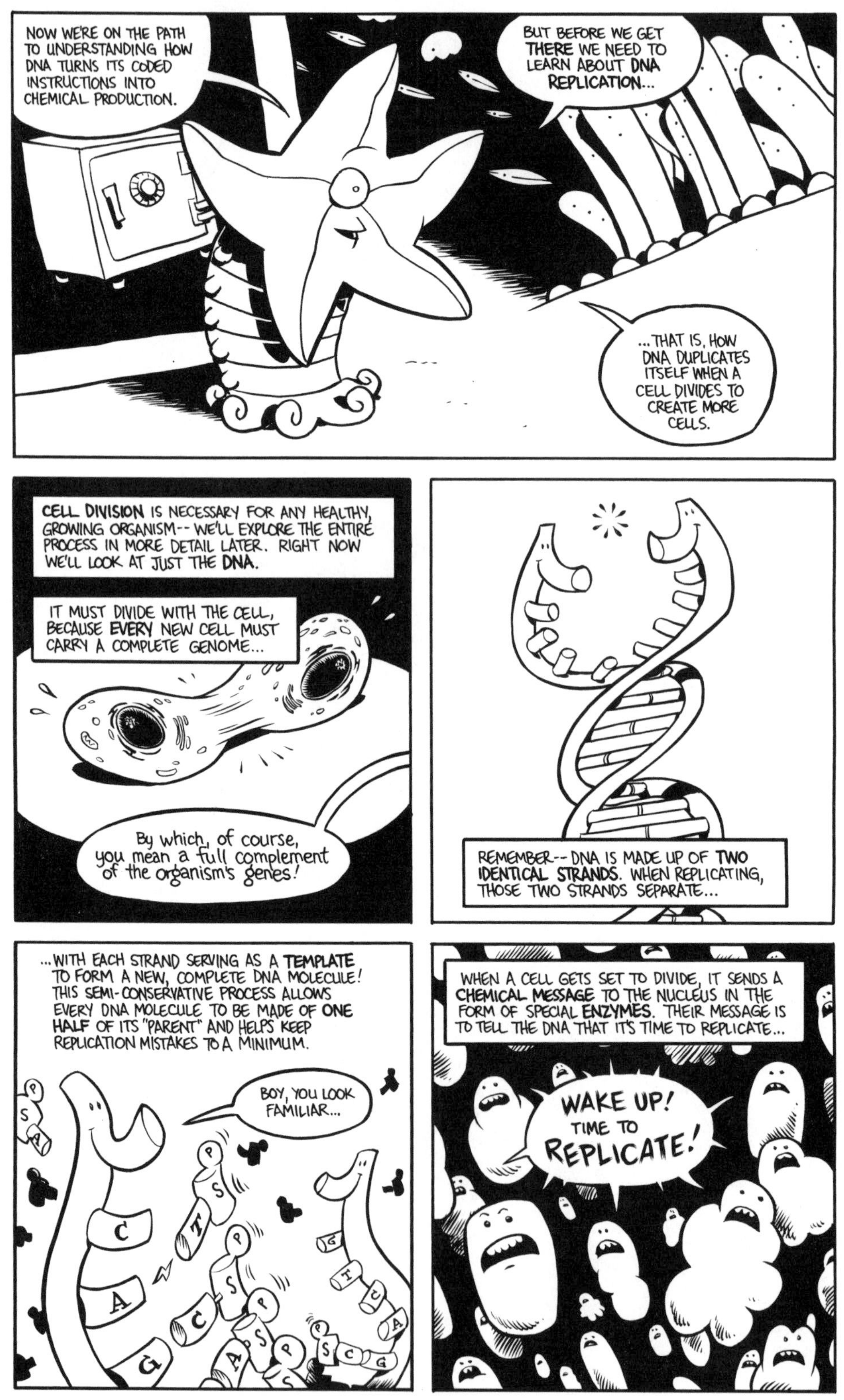
NOW WE'RE ON THE PATH TO UNDERSTANDING HOW DNA TURNS ITS CODED INSTRUCTIONS INTO CHEMICAL PRODUCTION.
BUT BEFORE WE GET THERE WE NEED TO LEARN ABOUT DNA REPLICATION...
...THAT IS, HOW DNA DUPLICATES ITSELF WHEN A CELL DIVIDES TO CREATE MORE CELLS.
CELL DIVISION IS NECESSARY FOR ANY HEALTHY, GROWING ORGANISM-- WE'LL EXPLORE THE ENTIRE PROCESS IN MORE DETAIL LATER. RIGHT NOW WE'LL LOOK AT JUST THE DNA.
IT MUST DIVIDE WITH THE CELL, BECAUSE EVERY NEW CELL MUST CARRY A COMPLETE GENOME...
By which, of course, you mean a full complement of the organism's genes!
REMEMBER-- DNA IS MADE UP OF TWO IDENTICAL STRANDS. WHEN REPLICATING, THOSE TWO STRANDS SEPARATE...
...WITH EACH STRAND SERVING AS A TEMPLATE TO FORM A NEW, COMPLETE DNA MOLECULE! THIS SEMI-CONSERVATIVE PROCESS ALLOWS EVERY DNA MOLECULE TO BE MADE OF ONE HALF OF ITS "PARENT" AND HELPS KEEP REPLICATION MISTAKES TO A MINIMUM.
BOY, YOU LOOK FAMILIAR...
WHEN A CELL GETS SET TO DIVIDE, IT SENDS A CHEMICAL MESSAGE TO THE NUCLEUS IN THE FORM OF SPECIAL ENZYMES. THEIR MESSAGE IS TO TELL THE DNA THAT IT'S TIME TO REPLICATE...
WAKE UP! TIME TO REPLICATE!

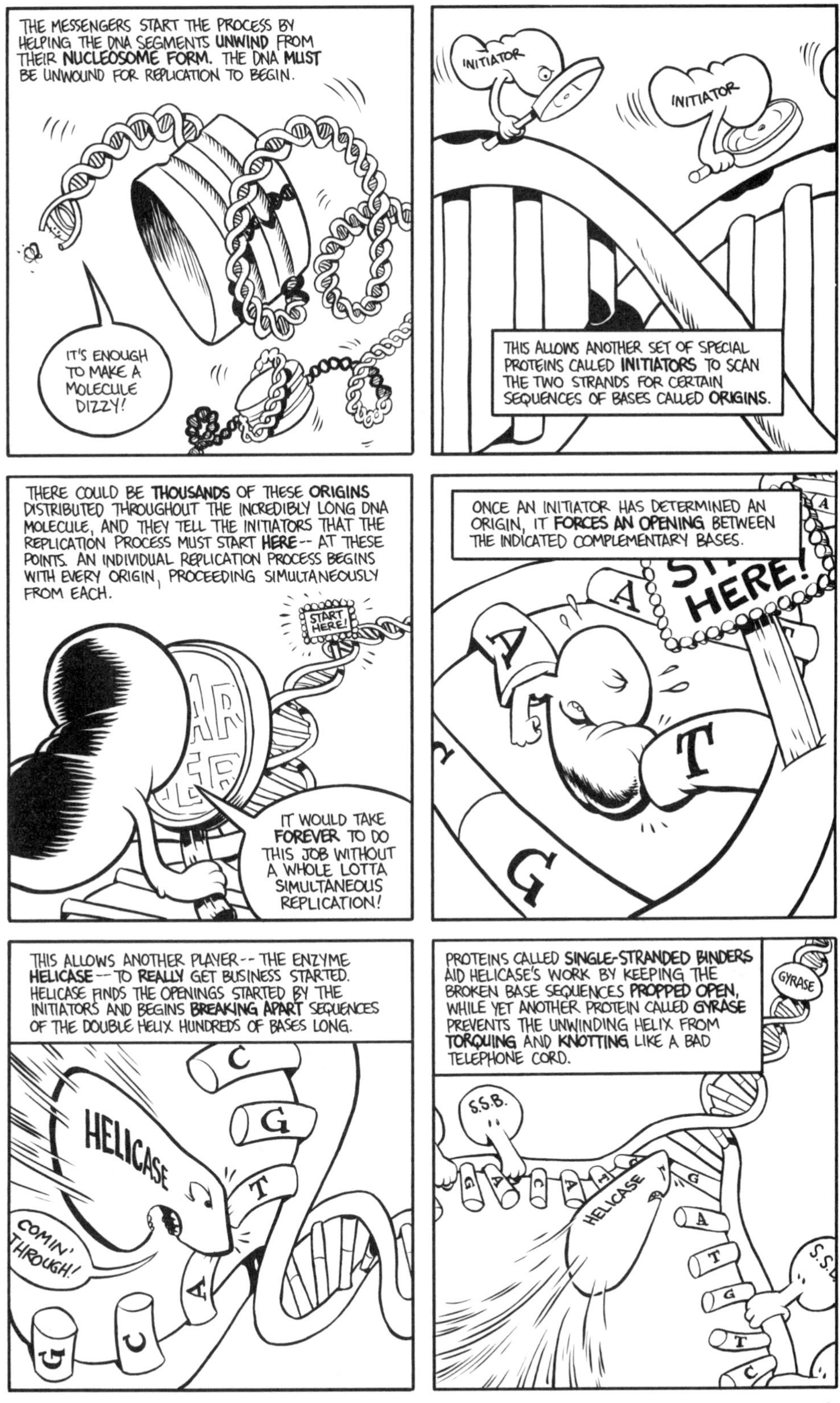
THE MESSENGERS START THE PROCESS BY HELPING THE DNA SEGMENTS UNWIND FROM THEIR NUCLEOSOME FORM. THE DNA MUST BE UNWOUND FOR REPLICATION TO BEGIN.
IT'S ENOUGH TO MAKE A MOLECULE DIZZY!
INITIATOR
INITIATOR
THIS ALLOWS ANOTHER SET OF SPECIAL PROTEINS CALLED INITIATORS TO SCAN THE TWO STRANDS FOR CERTAIN SEQUENCES OF BASES CALLED ORIGINS.
THERE COULD BE THOUSANDS OF THESE ORIGINS DISTRIBUTED THROUGHOUT THE INCREDIBLY LONG DNA MOLECULE, AND THEY TELL THE INITIATORS THAT THE REPLICATION PROCESS MUST START HERE-- AT THESE POINTS. AN INDIVIDUAL REPLICATION PROCESS BEGINS WITH EVERY ORIGIN, PROCEEDING SIMULTANEOUSLY FROM EACH.
START HERE!
IT WOULD TAKE FOREVER TO DO THIS JOB WITHOUT A WHOLE LOTTA SIMULTANEOUS REPLICATION!
ONCE AN INITIATOR HAS DETERMINED AN ORIGIN, IT FORCES AN OPENING BETWEEN THE INDICATED COMPLEMENTARY BASES.
HERE!
A
A
T
G
THIS ALLOWS ANOTHER PLAYER-- THE ENZYME HELICASE-- TO REALLY GET BUSINESS STARTED. HELICASE FINDS THE OPENINGS STARTED BY THE INITIATORS AND BEGINS BREAKING APART SEQUENCES OF THE DOUBLE HELIX HUNDREDS OF BASES LONG.
HELICASE
C
G
T
A
C
G
COMIN' THROUGH!
PROTEINS CALLED SINGLE-STRANDED BINDERS AID HELICASE'S WORK BY KEEPING THE BROKEN BASE SEQUENCES PROPPED OPEN, WHILE YET ANOTHER PROTEIN CALLED GYRASE PREVENTS THE UNWINDING HELIX FROM TORQUING AND KNOTTING LIKE A BAD TELEPHONE CORD.
GYRASE
S.S.B.
HELICASE
G
A
C
A
A
T
G
T
C

AS YOU CAN SEE HERE, HELICASE'S ATTACK FORMS TWO FORKS IN THE MOLECULE. REPLICATION CAN PROCEED DOWN BOTH OF THE TWO TEMPLATE STRANDS IN OPPOSITE DIRECTIONS.
NOW, TO GET THE ACTUAL REPLICATION ITSELF STARTED, A NUCLEOTIDE OF TEMPORARY RNA (RIBONUCLEIC ACID -- DNA's MORE INTERACTIVE COUSIN) COMES ALONG AND BONDS WITH THE BASE OF THE FIRST NUCLEOTIDE OPENED BY HELICASE.
HELLO THERE, SAILOR.
RNA NUCLEOTIDE
START HERE!
THIS BIT OF TEMPORARY RNA IS ESSENTIALLY PRIMING THE PUMP-- ONCE IT HAS GOT THE PROCESS ROLLING, ANOTHER ENZYME, DNA POLYMERASE, TAKES OVER AND DOES THE SERIOUS HEAVY LIFTING. DNA POLYMERASE MOVES DOWN THE OPENED CHAIN, IDENTIFIES THE BASES IN SEQUENCE, AND CONNECTS EACH WITH A DNA NUCLEOTIDE OF THE APPROPRIATE COMPLEMENTARY BASE.
THANK YOU, RNA-- I'LL TAKE IT FROM HERE.
DNA POLYMERASE
AND SO A COMPLETE, NEW DNA MOLECULE IS FORMED, IN ALL ITS DOUBLE-STRANDED GLORY, WITH HELICASE CONSTANTLY PLUNGING AHEAD, OPENING NEW BASES DOWN THE CHAIN...
...AND POLYMERASE FOLLOWING, ADDING COMPLEMENTARY NUCLEOTIDES, BASE BY BASE.
THIS PROCESS CONTINUES ON IN BOTH STRANDS OF THE TEMPLATE MOLECULE...
IF YOU THINK THIS IS COMPLEX, YOU SHOULD CHECK OUT OKAZAKI FRAGMENTS!
...UNTIL TWO NEW DNA MOLECULES ARE COMPLETE, EACH COMPRISED OF ONE ORIGINAL PARENT STRAND AND ONE NEW STRAND SYNTHESIZED FROM FREE-FLOATING NUCLEOTIDES.
ACTUALLY, WE'RE BOTH STILL A BIT FRAGMENTED, THANKS TO OUR MULTIPLE ORIGINS...

THERE ARE STILL SOME FINISHING TOUCHES THAT NEED TO BE ATTENDED TO. THOSE RNA PRIMER NUCLEOTIDES THAT GOT REPLICATION STARTED MUST BE REMOVED BY DNA POLYMERASE, AND REPLACED WITH ACTUAL DNA NUCLEOTIDES HAVING THE APPROPRIATE BASES...
TIME TO GO.
ULP!
DNA POLYMERASE
...AND THEN ALL THE FRAGMENTED SECTIONS OF THE NEW DNA MOLECULE--EACH STARTED AT ITS OWN ORIGIN--MUST BE JOINED TOGETHER BY THE ENZYME LIGASE, LINKING PHOSPHATES TO SUGARS WITH THEIR POWERFUL BOND.
EVERYBODY GET IN LINE!
LIGASE
THAT'S A LOT OF ENZYMES AND A LOT OF STEPS TO COMPLETE, AND REPLICATION TRANSPIRES QUICKLY. IN HUMANS, 2,000 BASES ARE REPLICATED PER MINUTE! GIVEN THE HIGH DEGREE OF DIFFICULTY, IT'S NO WONDER THAT MISTAKES--ALTHOUGH AMAZINGLY INFREQUENT--DO OCCUR.
HEY, DO YOU WANT IT DONE FAST, OR RIGHT?
WE DO PRETTY WELL, CONSIDERING THE CIRCUMSTANCES.
DNA POLYMERASE
PROOFREADING DNA
AHA! YOU LOOK ABSOLUTELY REPELLANT!
BASES DO OCCASIONALLY GET MISMATCHED. THE RESULT IS A BAD BOND--A BASE PAIR THAT DOES NOT JOIN PROPERLY. THIS ACTION CREATES A "BUMP"--A MALFORMATION IN THE MOLECULAR STRUCTURE. PROOFREADING DNA POLYMERASE IDENTIFIES THE MISMATCHED BASE AND REPLACES IT WITH THE APPROPRIATE COMPLEMENT.
IT NEVER HURTS TO CHECK AGAIN...
ANY MISTAKES THAT GET BY THE PROOFREADING DNA POLYMERASE ARE CAUGHT BY YET ANOTHER SYSTEM OF SPECIALIZED ENZYMES THAT EFFECT MISMATCH REPAIR AFTER REPLICATION IS COMPLETE.
SUCH IS THE NEED FOR TOTAL ACCURACY OF REPLICATION IN DNA THAT WE SEE SO MANY SEEMINGLY REDUNDANT FAIL-SAFES BUILT INTO THE SYSTEM!

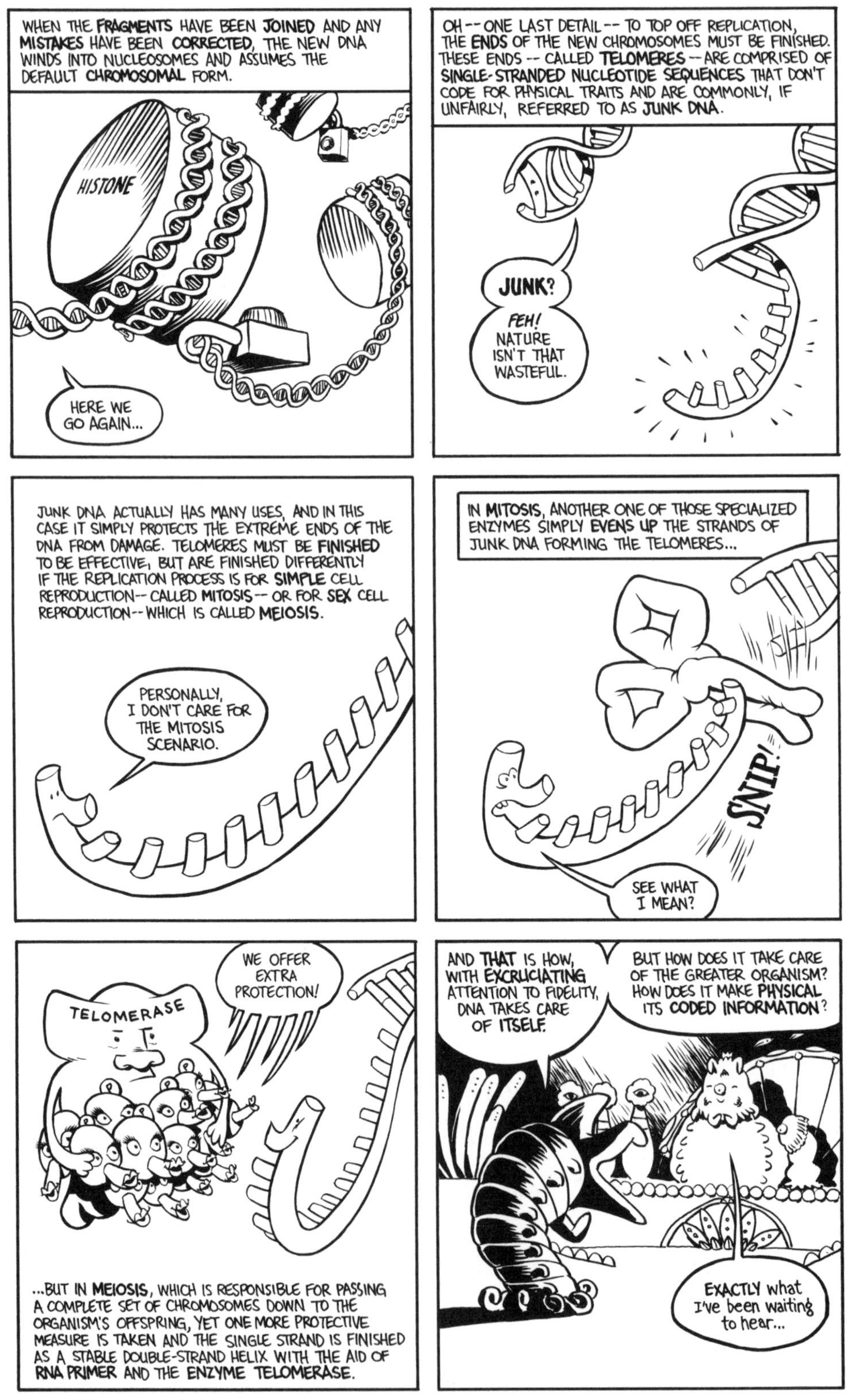
WHEN THE FRAGMENTS HAVE BEEN JOINED AND ANY MISTAKES HAVE BEEN CORRECTED, THE NEW DNA WINDS INTO NUCLEOSOMES AND ASSUMES THE DEFAULT CHROMOSOMAL FORM.
HISTONE
HERE WE GO AGAIN...
OH -- ONE LAST DETAIL -- TO TOP OFF REPLICATION, THE ENDS OF THE NEW CHROMOSOMES MUST BE FINISHED. THESE ENDS -- CALLED TELOMERES -- ARE COMPRISED OF SINGLE-STRANDED NUCLEOTIDE SEQUENCES THAT DON'T CODE FOR PHYSICAL TRAITS AND ARE COMMONLY, IF UNFAIRLY, REFERRED TO AS JUNK DNA.
JUNK?
FEH! NATURE ISN'T THAT WASTEFUL.
JUNK DNA ACTUALLY HAS MANY USES, AND IN THIS CASE IT SIMPLY PROTECTS THE EXTREME ENDS OF THE DNA FROM DAMAGE. TELOMERES MUST BE FINISHED TO BE EFFECTIVE, BUT ARE FINISHED DIFFERENTLY IF THE REPLICATION PROCESS IS FOR SIMPLE CELL REPRODUCTION -- CALLED MITOSIS -- OR FOR SEX CELL REPRODUCTION -- WHICH IS CALLED MEIOSIS.
PERSONALLY, I DON'T CARE FOR THE MITOSIS SCENARIO.
IN MITOSIS, ANOTHER ONE OF THOSE SPECIALIZED ENZYMES SIMPLY EVENS UP THE STRANDS OF JUNK DNA FORMING THE TELOMERES...
SNIP!
SEE WHAT I MEAN?
TELOMERASE
WE OFFER EXTRA PROTECTION!
...BUT IN MEIOSIS, WHICH IS RESPONSIBLE FOR PASSING A COMPLETE SET OF CHROMOSOMES DOWN TO THE ORGANISM'S OFFSPRING, YET ONE MORE PROTECTIVE MEASURE IS TAKEN AND THE SINGLE STRAND IS FINISHED AS A STABLE DOUBLE-STRAND HELIX WITH THE AID OF RNA PRIMER AND THE ENZYME TELOMERASE.
AND THAT IS HOW, WITH EXCRUCIATING ATTENTION TO FIDELITY, DNA TAKES CARE OF ITSELF.
BUT HOW DOES IT TAKE CARE OF THE GREATER ORGANISM? HOW DOES IT MAKE PHYSICAL ITS CODED INFORMATION?
EXACTLY what I've been waiting to hear...

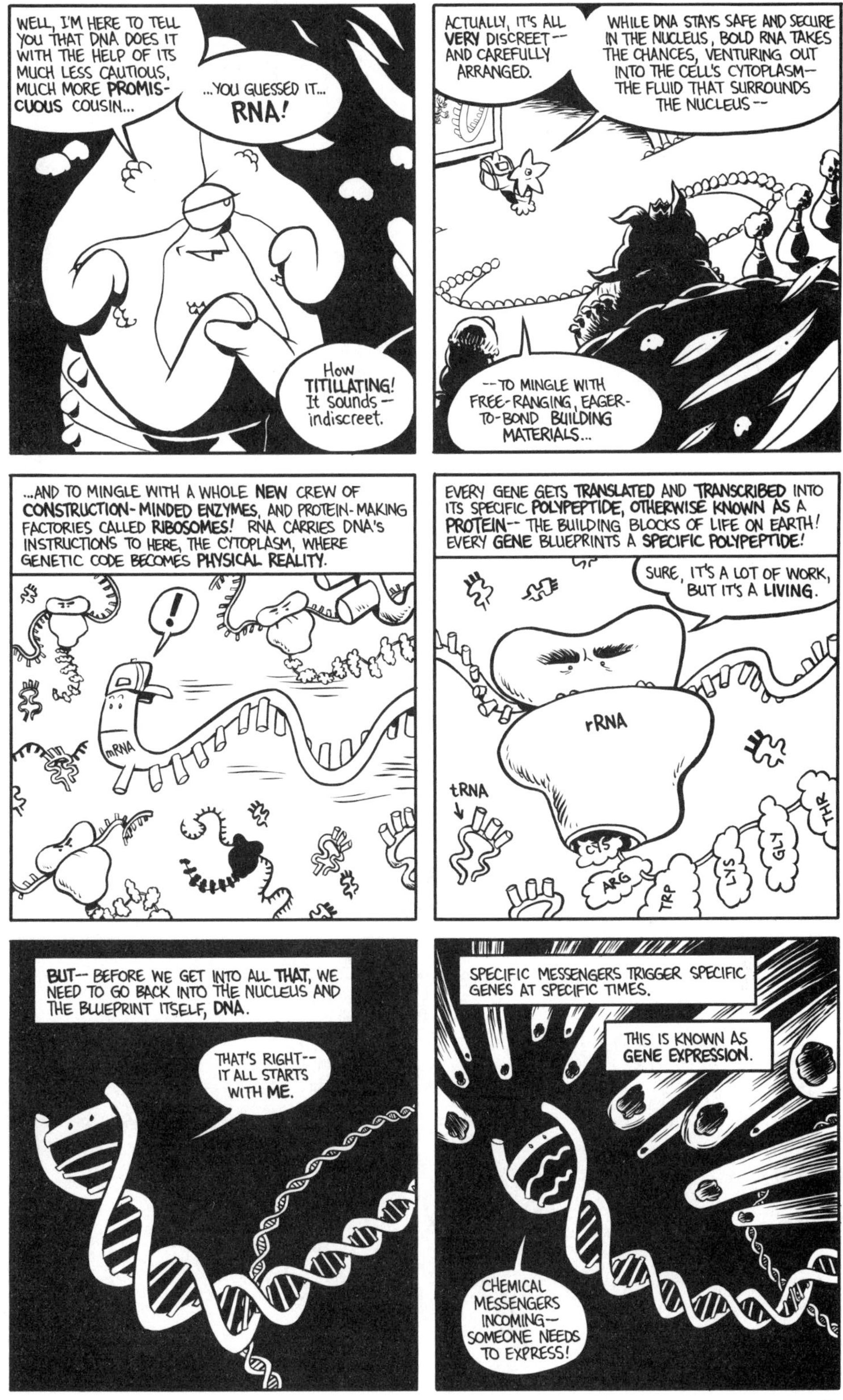
WELL, I'M HERE TO TELL YOU THAT DNA DOES IT WITH THE HELP OF ITS MUCH LESS CAUTIOUS, MUCH MORE PROMISCUOUS COUSIN...
...YOU GUESSED IT... RNA!
How TITILLATING! It sounds--indiscreet.
ACTUALLY, IT'S ALL VERY DISCREET--AND CAREFULLY ARRANGED.
WHILE DNA STAYS SAFE AND SECURE IN THE NUCLEUS, BOLD RNA TAKES THE CHANCES, VENTURING OUT INTO THE CELL'S CYTOPLASM— THE FLUID THAT SURROUNDS THE NUCLEUS--
--TO MINGLE WITH FREE-RANGING, EAGER-TO-BOND BUILDING MATERIALS...
...AND TO MINGLE WITH A WHOLE NEW CREW OF CONSTRUCTION-MINDED ENZYMES, AND PROTEIN-MAKING FACTORIES CALLED RIBOSOMES! RNA CARRIES DNA'S INSTRUCTIONS TO HERE, THE CYTOPLASM, WHERE GENETIC CODE BECOMES PHYSICAL REALITY.
!
mRNA
EVERY GENE GETS TRANSLATED AND TRANSCRIBED INTO ITS SPECIFIC POLYPEPTIDE, OTHERWISE KNOWN AS A PROTEIN-- THE BUILDING BLOCKS OF LIFE ON EARTH! EVERY GENE BLUEPRINTS A SPECIFIC POLYPEPTIDE!
SURE, IT'S A LOT OF WORK, BUT IT'S A LIVING.
rRNA
tRNA
CYS
ARG
TRP
LYS
GLY
THR
BUT-- BEFORE WE GET INTO ALL THAT, WE NEED TO GO BACK INTO THE NUCLEUS AND THE BLUEPRINT ITSELF, DNA.
THAT'S RIGHT-- IT ALL STARTS WITH ME.
SPECIFIC MESSENGERS TRIGGER SPECIFIC GENES AT SPECIFIC TIMES.
THIS IS KNOWN AS GENE EXPRESSION.
CHEMICAL MESSENGERS INCOMING— SOMEONE NEEDS TO EXPRESS!

DNA from a HUMAN PERSPECTIVE
PART 2
DNA MAY HAVE BEEN ACCEPTED AS THE GENETIC MATERIAL BY THE EARLY 1950s, BUT ITS STRUCTURE -- THE SHAPE THAT ALLOWS IT TO DO ITS REMARKABLE WORK-- WAS STILL UNKNOWN.
JAMES WATSON
JAMES WATSON AND FRANCIS CRICK BEGAN A RESEARCH PARTNERSHIP AT CAMBRIDGE UNIVERSITY IN 1951, WITH THE GOAL OF DISCOVERING THAT STRUCTURE. WATSON SPECIALIZED IN X-RAY PHOTOGRAPHY, AND HIS RESULTS WERE LEADING THE TWO TO BELIEVE THAT DNA'S OVERALL SHAPE WAS THAT OF A HELIX.
NOT FAR AWAY, AT KING'S COLLEGE, LONDON, ROSALIND FRANKLIN BEGAN SIMILAR RESEARCH THAT SAME YEAR WITH MAURICE WILKINS. HER EXCELLENT X-RAY PHOTOGRAPHY OF DNA ALSO INDICATED A POSSIBLE HELIX SHAPE, ALTHOUGH SHE WAS CAUTIOUS ABOUT PUBLICLY ANNOUNCING HER FINDINGS.
FRANCIS CRICK
WATSON AND CRICK INITIALLY THOUGHT THE STRUCTURE OF DNA PLACED THE SUGAR/PHOSPHATE BACKBONE AT THE CENTER OF THE MACROMOLECULE. BUT, AFTER ATTENDING A LECTURE BY FRANKLIN FEATURING HER RESEARCH INDICATING THE OPPOSITE, THEY WERE CONVINCED TO RETHINK THEIR MODEL AND CORRECTLY REPOSITIONED THE SUGAR/PHOSPHATE COMPONENT OF DNA AS A FIRM BUT FLEXIBLE OUTER SKIN, WITH THE FOUR BASES ATTACHING ON THE INSIDE OF IT.
WILKINS ALSO SUPPOSEDLY ALLOWED WATSON TO VIEW FRANKLIN'S UNPUBLISHED RESEARCH PAPERS. THIS UNETHICAL ACTION WOULD CERTAINLY HAVE FURTHERED WATSON AND CRICK'S RACE TO DISCOVER DNA'S STRUCTURE FIRST. THE CHARGE REMAINS CONTROVERSIAL.
ROSALIND FRANKLIN
WATSON AND CRICK APPLIED THEIR RESEARCH-GUIDED THEORIES TO MODELS CONSTRUCTED OF CARDBOARD AND PASTE, REARRANGING MOLECULAR PLACEMENTS UNTIL THEY EVENTUALLY LOCKED TOGETHER IN LOGICAL ATOMIC BONDS. TOO MANY PIECES OF THE PUZZLE FIT FOR THEIR MODEL TO BE WRONG-- THE NOW-FAMILIAR DOUBLE-HELIX STRUCTURE OF DNA WAS REVEALED FOR THE FIRST TIME!
WATSON AND CRICK, ALONG WITH WILKINS, WERE AWARDED THE NOBEL PRIZE FOR MEDICINE IN 1962.
FRANKLIN, WHO HAD CONTRIBUTED MUCH TO THE DISCOVERY, WAS NOT ELIGIBLE FOR THE DISTINCTION. SHE DIED IN 1958 OF CANCER, A GENETICALLY RELATED DISEASE THAT SOMEDAY MAY BE BANISHED AS OUR UNDERSTANDING OF DNA GROWS.

REMEMBER--ALMOST EVERY TYPE OF CELL OF EVERY ORGANISM CARRIES THAT ORGANISM'S COMPLETE GENOME--ALL ITS GENETIC INSTRUCTIONS. SO, IF THESE CELLS CARRY ALL GENES...
RED BLOOD CELLS ARE PRIVILEGED CHARACTERS--THEY DON'T BOTHER TO CARRY ALL THOSE GENETIC INSTRUCTIONS.
...WHY DOESN'T SKIN SPROUT EYEBALLS--WHY DON'T LIVERS GROW TOES? WHY DO THE APPROPRIATE CHARACTERISTICS DEVELOP ONLY IN THE APPROPRIATE ORGANS AT THE APPROPRIATE TIME?
BECAUSE GENES ARE CALLED ON TO EXPRESS--TO BECOME ACTIVE--ONLY IN CERTAIN CELLS AT CERTAIN TIMES. THEY ARE SWITCHED ON OR OFF AS THEY ARE NEEDED.
GENE
ON
IN MOST CELLS, THE DEFAULT SETTING IS "OFF." WHY? BECAUSE THE UNRESTRICTED GROWTH OF A PARTICULAR PHYSICAL CHARACTERISTIC CAN BE BAD.
NO KIDDING...
CANCER IS THE RESULT OF AN UNCONTROLLED GENE EXPRESSION.
IN THE EARLY STAGES OF AN ORGANISM'S DEVELOPMENT, MOST GENES BECOME TISSUE SPECIFIC. THAT MEANS THAT A SPECIFIC CELL'S DNA PERMANENTLY TURNS OFF ALL ITS GENES THAT DO NOT CODE FOR PHYSICAL TRAITS SPECIFICALLY NEEDED FOR THAT CELL.
BUT BEFORE THAT, ALL CELLS ARE TOTIPOTENT--WE CAN BECOME ANY TISSUE NEEDED!
BUT WHAT ARE THE FACTORS THAT ACTIVATE A CELL'S "TISSUE APPROPRIATE" GENES WHEN THEY ARE NEEDED?
WE'VE GOT INCOMING!
THERE ARE SEVERAL, SUCH AS...

...ENVIRONMENTAL FACTORS--CONDITIONS GENERATED OUTSIDE THE ORGANISM. MOST IMPORTANT, PERHAPS, ARE THE EFFECTS HEAT AND LIGHT HAVE ON GENE EXPRESSION.
MAKES ME TURN EXPRESSLY IN THIS DIRECTION.
ALSO, CERTAIN GENES ARE PRESET TO ACTIVATE AT CERTAIN STAGES OF AN ORGANISM'S DEVELOPMENT. THEY RECEIVE THEIR CUES FROM CHANGES WITHIN THE CELL OR FROM NEIGHBORING CELLS.
PASS THE WORD--TIME TO EXPRESS!
AND THEN THERE ARE HORMONES--COMPLEX CHEMICALS PRODUCED BY THE BRAIN AND VARIOUS GLANDS THAT RUSH THROUGH THE CIRCULATORY SYSTEM, BRINGING THE INSTRUCTIONS TO ACTIVATE AND DEACTIVATE GENES.
MD
WE DO MORE THAN JUST GET ADOLESCENTS ALL HOT AND BOTHERED...
Oh, my...
WHEN A GENE DOES GET SWITCHED ON, A PROCESS CALLED TRANSCRIPTION BEGINS--THAT GENE BECOMES "TRANSCRIBED," OR DUPLICATED, INTO A MESSENGER RNA MOLECULE.
A RECORD OF THE GENE'S BASE-PAIR CODE--THE EXACT SEQUENCE OF BASES--IS BUILT OUT OF RNA.
REMEMBER: RNA IS A NUCLEIC ACID LIKE ITS COUSIN DNA, BUT ITS SUGAR--RIBOSE--CARRIES ONE MORE OXYGEN ATOM THAN DNA'S DOES. THAT DOESN'T SEEM LIKE A BIG DIFFERENCE, BUT THE EXTRA O MAKES RNA MUCH FRIENDLIER, MUCH MORE PRONE TO CANOODLING WITH OTHER MOLECULES THAN DNA.
RIBOSE
DEOXYRIBOSE
AND THAT'S AN IMPORTANT DISTINCTION, AS WE'LL SEE.
URACIL
THYMINE
ANOTHER IMPORTANT DISTINCTION: RNA'S BASES ARE SLIGHTLY DIFFERENT FROM DNA'S. WHILE BOTH EMPLOY GUANINE, CYTOSINE, AND ADENINE, RNA USES URACIL (U) INSTEAD OF THYMINE.

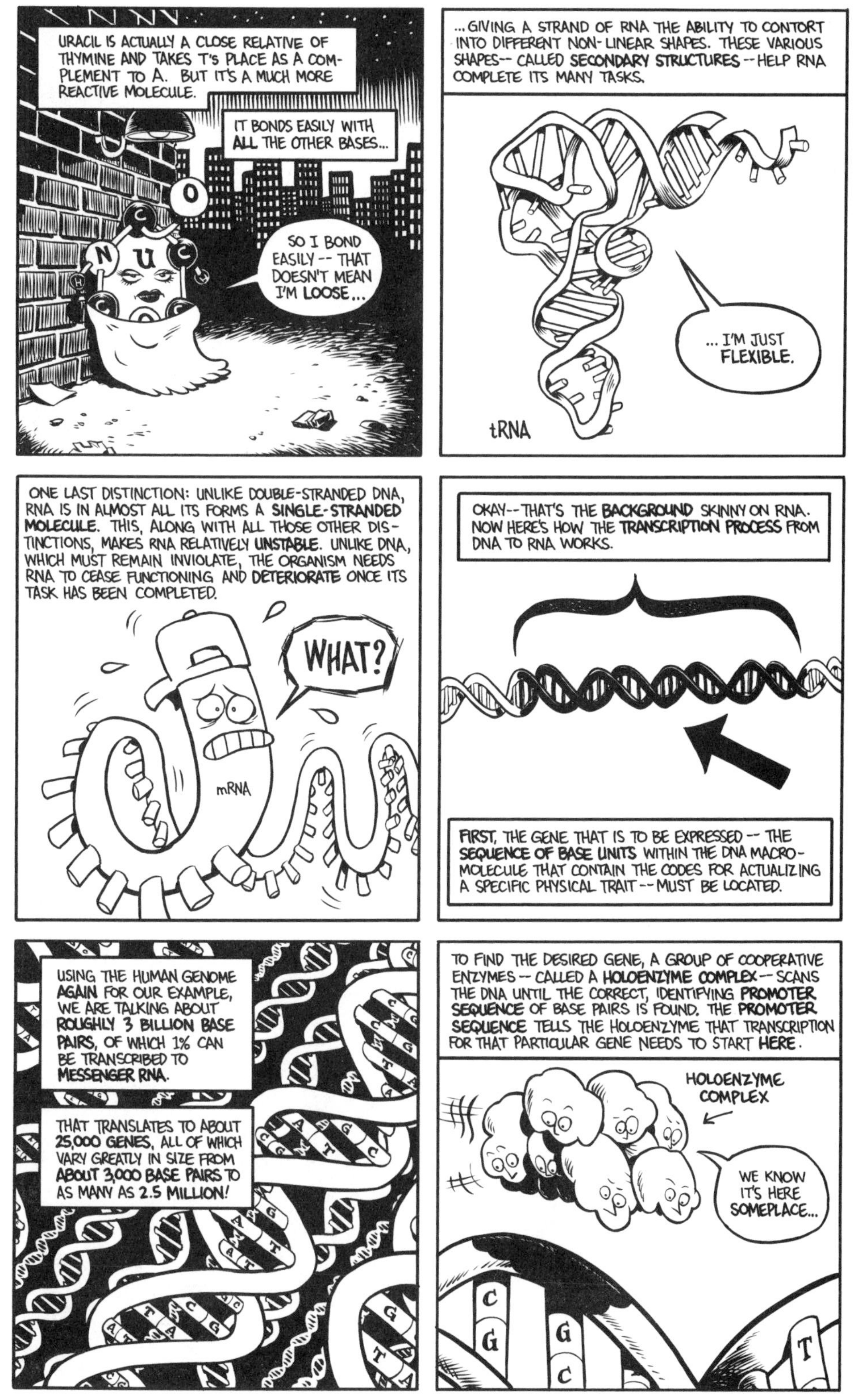
URACIL IS ACTUALLY A CLOSE RELATIVE OF THYMINE AND TAKES T'S PLACE AS A COMPLEMENT TO A. BUT IT'S A MUCH MORE REACTIVE MOLECULE.
IT BONDS EASILY WITH ALL THE OTHER BASES...
O
C
N
U
C
H
H
SO I BOND EASILY -- THAT DOESN'T MEAN I'M LOOSE...
...GIVING A STRAND OF RNA THE ABILITY TO CONTORT INTO DIFFERENT NON-LINEAR SHAPES. THESE VARIOUS SHAPES-- CALLED SECONDARY STRUCTURES --HELP RNA COMPLETE ITS MANY TASKS.
... I'M JUST FLEXIBLE.
tRNA
ONE LAST DISTINCTION: UNLIKE DOUBLE-STRANDED DNA, RNA IS IN ALMOST ALL ITS FORMS A SINGLE-STRANDED MOLECULE. THIS, ALONG WITH ALL THOSE OTHER DISTINCTIONS, MAKES RNA RELATIVELY UNSTABLE. UNLIKE DNA, WHICH MUST REMAIN INVIOLATE, THE ORGANISM NEEDS RNA TO CEASE FUNCTIONING AND DETERIORATE ONCE ITS TASK HAS BEEN COMPLETED.
WHAT?
mRNA
OKAY--THAT'S THE BACKGROUND SKINNY ON RNA. NOW HERE'S HOW THE TRANSCRIPTION PROCESS FROM DNA TO RNA WORKS.
FIRST, THE GENE THAT IS TO BE EXPRESSED -- THE SEQUENCE OF BASE UNITS WITHIN THE DNA MACROMOLECULE THAT CONTAIN THE CODES FOR ACTUALIZING A SPECIFIC PHYSICAL TRAIT--MUST BE LOCATED.
USING THE HUMAN GENOME AGAIN FOR OUR EXAMPLE, WE ARE TALKING ABOUT ROUGHLY 3 BILLION BASE PAIRS, OF WHICH 1% CAN BE TRANSCRIBED TO MESSENGER RNA.
THAT TRANSLATES TO ABOUT 25,000 GENES, ALL OF WHICH VARY GREATLY IN SIZE FROM ABOUT 3,000 BASE PAIRS TO AS MANY AS 2.5 MILLION!
TO FIND THE DESIRED GENE, A GROUP OF COOPERATIVE ENZYMES -- CALLED A HOLOENZYME COMPLEX -- SCANS THE DNA UNTIL THE CORRECT, IDENTIFYING PROMOTER SEQUENCE OF BASE PAIRS IS FOUND. THE PROMOTER SEQUENCE TELLS THE HOLOENZYME THAT TRANSCRIPTION FOR THAT PARTICULAR GENE NEEDS TO START HERE.
HOLOENZYME COMPLEX
WE KNOW IT'S HERE SOMEPLACE...
C
G
G
C
T

THE PROMOTER SEQUENCE -- OFTEN CODED "TATA," AS SHOWN HERE -- IS ALWAYS LOCATED ON THE DNA STRAND NOT USED AS THE TEMPLATE FOR TRANSCRIPTION TO MESSENGER RNA (mRNA). IT TELLS THE ENZYMES TO TRANSCRIBE FROM THE OPPOSITE STRAND.
THE PROMOTER ALSO CONTROLS HOW MANY TIMES THE GENE IS TRANSCRIBED TO mRNA, BASED ON HOW MUCH PHYSICAL MATERIAL IS REQUIRED -- HOW MANY NEW COPIES OF THAT PARTICULAR GENE THE ORGANISM NEEDS.
NUTHIN' STARTS WITHOUT MY SAY-SO.
ONCE THE PROMOTER SEQUENCE HAS BEEN CORRECTLY IDENTIFIED, THE HOLOENZYME COMPLEX BINDS TO IT AND SIGNALS RNA POLYMERASE TO STAND BY FOR ACTION!
BUT BE GENTLE, ALL RIGHT?!
WHEN THE PROMOTER SEQUENCE SIGNALS "READY!" THE HOLOENZYME COMPLEX MAKES A TINY BREAK IN THE FIRST NUCLEOTIDES OF THE GENE, ALLOWING RNA POLYMERASE TO SEE PAST THE SUGAR-PHOSPHATES TO THE BASE PAIRS.
RNA POLYMERASE
GOIN' IN!
ONCE RNA POLYMERASE CAN GET AT THE BASE PAIRS, IT BEGINS BREAKING APART A STRING OF BASE PAIRS ABOUT 20 PAIRS LONG -- THIS IS CALLED THE TRANSCRIPTION BUBBLE. NOW mRNA SYNTHESIS CAN BEGIN.
RNA POLYMERASE
COMIN' THROUGH!
UNLIKE THE DNA TRANSLATION PROCESS, RNA TRANSCRIPTION DOESN'T NEED A PRIMER TO GET ITS SHOW ON THE ROAD. IT JUST READS THE FIRST BASE OF THE TRANSCRIPTION BUBBLE AND STARTS PULLING IN FREE-FLOATING NUCLEOTIDES CARRYING THE APPROPRIATE COMPLEMENTARY BASES.
FREE RNA NUCLEOTIDES

THE NUCLEOTIDES FORM THEIR PHOSPHATE-TO-SUGAR BONDS AND THE PROCESS CONTINUES DOWN THE GENE. RNA POLYMERASE OPENS THE HELIX--AND CONTINUES THE TRANSCRIPTION BUBBLE AND SYNTHESIZING MORE mRNA.
RNA POLYMERASE
SUGAR
SUGAR
SUGAR
SUGAR
P
G
C
U
A
A
T
C
G
AS THE mRNA STRAND GROWS, IT IS PUSHED AWAY FROM ITS TEMPLATE, AND THE DNA SNAPS CLOSED AS THE TRANSCRIPTION BUBBLE MOVES ON. THIS ONGOING TRANSCRIPTION PROCESS IS CALLED **ELONGATION**.
WHEN THE GENE IS **COMPLETELY** TRANSCRIBED, RNA POLYMERASE HITS A **TERMINATOR SEQUENCE OF BASES** THAT TELLS IT TO STOP AND TO **DETACH** FROM THE DNA TEMPLATE. THE DNA MOLECULE SNAPS BACK TO ITS USUAL DOUBLE-HELIX SHAPE, ITS JOB DONE!
RNA POLYMERASE
STOP!
END OF THE LINE FOR YOU, BUDDY.
WORK CONTINUES, HOWEVER, ON OUR NEW mRNA STRAND. LIKE GENOMES IN GENERAL, THERE ARE SEQUENCES OF BASES WITHIN TRANSCRIPTION UNITS THAT DON'T CODE FOR **PHENOTYPE** -- PHYSICAL TRAITS, THAT IS. THIS **"JUNK"** IS REFERRED TO AS **INTRONS**, A FANCY NAME MEANING "INTERVENING SEQUENCES."
JUNK -- A CONVENIENT CATCH-ALL SLUR IF I EVER HEARD ONE! WE INTRONS MAY BE MORE IMPORTANT THAN YOU THINK!
A CLEVERLY NAMED ENZYME CALLED A **SPLICEOSOME** RECOGNIZES THESE INTRONS AND EDITS THEM OUT. THEN IT **SPLICES** TOGETHER THE REMAINING **EXON** SEQUENCES (THOSE THAT **DO** CODE FOR A PHENOTYPE) INTO A SEAMLESS, WASTE-FREE **MESSAGE DELIVERY SYSTEM** OF DNA'S INSTRUCTIONS!
THERE'S NO PLACE HERE FOR THOSE WHO DON'T TRANSCRIBE TO PROTEIN!
EXONS
INTRONS
TO FINISH OFF OUR STREAMLINED mRNA STRAND, A **CAP OF GUANINE** IS ADDED TO ITS LEADING END, WHILE **ADENINES** ARE STUCK ON THE TAIL. THESE PROTECT THE UNSTABLE MOLECULE FROM DECAYING **TOO** QUICKLY...
G
mRNA
A
WHAT THE WELL-DRESSED MESSENGER WILL BE WEARING!

...BECAUSE mRNA NOW HAS ITS MISSION TO COMPLETE. THAT INVOLVES CONVEYING THE INSTRUCTIONS IT HAS TRANSCRIBED FROM DNA OUT OF THE NUCLEUS...
NUCLEUS
mRNA
I GOTTA GET OUT OF THIS PLACE...
...IF IT'S THE LAST THING I EVER DO...
...AND INTO THE CYTOPLASM AND A WORLD OF MOLECULAR RAW MATERIAL AND CHEMICAL FACTORIES, ALL WAITING TO CREATE NEW PROTEIN -- THE CONSTRUCTION MATERIAL NECESSARY TO ALL LIFE FUNCTIONS!
!
mRNA
THERE ARE LOTS OF IMPORTANT THINGS FLOATING AROUND IN THE CYTOPLASM, BUT OUR IMMEDIATE CONCERN IS RESTRICTED TO AMINO ACIDS AND TWO OTHER RNA VARIANTS: RIBOSOMAL RNA, CLEVERLY REFERRED TO AS rRNA, AND TRANSFER RNA, WHICH IS CALLED -- YUP, YOU GUESSED IT! -- tRNA.
YEAH, THAT'S RIGHT-- WE'RE RELATED. WHAT ABOUT IT?
RIBOSOMAL RNA
TRANSFER RNA
AMINO ACIDS ARE THE MOLECULAR COMPONENTS THAT FORM THOSE LONG POLYPEPTIDE CHAINS -- PROTEINS AND ENZYMES. POLYPEPTIDES CAN BE 50 TO 1,000 AMINO ACIDS IN LENGTH, AND THAT ALLOWS FOR A GREAT VARIETY OF PROTEINS, INDEED.
ARGININE
TYROSINE
GLUTAMINE
ALANINE
DNA'S INSTRUCTIONS FOR ASSEMBLING THE CYTOPLASM'S LOOSE AMINO ACIDS INTO A SPECIFIC PROTEIN HAVE BEEN ENCODED IN mRNA BY TRANSCRIPTION.
NOW THAT CODE MUST BE TRANSLATED TO THE LANGUAGE OF THE PROTEINS.
Hmmm...
RNA to PROTEIN DICTIONARY
P
P
S
S
S
A
C
G
CODON
THE SECRET OF THE CODE CARRIED BY mRNA IS THAT THE BASES MUST BE READ THREE AT A TIME. THIS TRIPLET IS CALLED A CODON. THE FOUR DIFFERENT BASES (A, C, G, AND U), READ IN COMBINATIONS OF ANY THREE, YIELD A TOTAL OF 64 POSSIBLE CODONS.

THESE 64 DIFFERENT CODONS CODE FOR THE 20 DIFFERENT VARIETIES OF AMINO ACID. THERE'S A GOOD DEAL OF REDUNDANCY HERE, WITH A NUMBER OF BASE COMBINATIONS CODING FOR THE SAME PROTEIN. ACTUALLY, **61** OF THE **CODONS** CODE FOR THE **20 TYPES OF AMINO ACIDS**, WHILE THE REMAINING 3 ARE SIGNALS TO **"STOP"** THE TRANSLATION PROCESS.

FIRST LETTER ↓	SECOND LETTER: U	C	A	G	THIRD LETTER ↓
U	PHENYLALANINE PHENYLALANINE LEUCINE LEUCINE	SERINE SERINE SERINE SERINE	TYROSINE TYROSINE **STOP** **STOP**	CYSTEINE CYSTEINE **STOP** TRYPTOPHAN	U C A G
C	LEUCINE LEUCINE LEUCINE LEUCINE	PROLINE PROLINE PROLINE PROLINE	HISTIDINE HISTIDINE GLUTAMINE GLUTAMINE	ARGININE ARGININE ARGININE ARGININE	U C A G
A	ISOLEUCINE ISOLEUCINE ISOLEUCINE METHIONINE **& START**	THREONINE THREONINE THREONINE THREONINE	ASPARAGINE ASPARAGINE LYSINE LYSINE	SERINE SERINE ARGININE ARGININE	U C A G
G	VALINE VALINE VALINE VALINE	ALANINE ALANINE ALANINE ALANINE	ASPARTATE ASPARTATE GLUTAMATE GLUTAMATE	GLYCINE GLYCINE GLYCINE GLYCINE	U C A G

SO WHAT **READS** THE CODONS?

RIBOSOMES -- THE rRNA FACTORIES THAT **PRODUCE** PROTEINS -- DO! RIBOSOMES CAN READ AND WORK WITH ANY mRNA STRAND, AND THEY COME IN **TWO** PARTS -- A **SMALL** AND A **LARGE SUBUNIT**.

THE DIVISION OF LABOR!

DNA from a HUMAN PERSPECTIVE
PART 2
JAMES WATSON
DNA MAY HAVE BEEN ACCEPTED AS THE GENETIC MATERIAL BY THE EARLY 1950s, BUT ITS STRUCTURE -- THE SHAPE THAT ALLOWS IT TO DO ITS REMARKABLE WORK-- WAS STILL UNKNOWN.
JAMES WATSON AND FRANCIS CRICK BEGAN A RESEARCH PARTNERSHIP AT CAMBRIDGE UNIVERSITY IN 1951, WITH THE GOAL OF DISCOVERING THAT STRUCTURE. WATSON SPECIALIZED IN X-RAY PHOTOGRAPHY, AND HIS RESULTS WERE LEADING THE TWO TO BELIEVE THAT DNA'S OVERALL SHAPE WAS THAT OF A HELIX.
NOT FAR AWAY, AT KING'S COLLEGE, LONDON, ROSALIND FRANKLIN BEGAN SIMILAR RESEARCH THAT SAME YEAR WITH MAURICE WILKINS. HER EXCELLENT X-RAY PHOTOGRAPHY OF DNA ALSO INDICATED A POSSIBLE HELIX SHAPE, ALTHOUGH SHE WAS CAUTIOUS ABOUT PUBLICLY ANNOUNCING HER FINDINGS.
FRANCIS CRICK
WATSON AND CRICK INITIALLY THOUGHT THE STRUCTURE OF DNA PLACED THE SUGAR/PHOSPHATE BACKBONE AT THE CENTER OF THE MACROMOLECULE. BUT, AFTER ATTENDING A LECTURE BY FRANKLIN FEATURING HER RESEARCH INDICATING THE OPPOSITE, THEY WERE CONVINCED TO RETHINK THEIR MODEL AND CORRECTLY REPOSITIONED THE SUGAR/PHOSPHATE COMPONENT OF DNA AS A FIRM BUT FLEXIBLE OUTER SKIN, WITH THE FOUR BASES ATTACHING ON THE INSIDE OF IT.
WILKINS ALSO SUPPOSEDLY ALLOWED WATSON TO VIEW FRANKLIN'S UNPUBLISHED RESEARCH PAPERS. THIS UNETHICAL ACTION WOULD CERTAINLY HAVE FURTHERED WATSON AND CRICK'S RACE TO DISCOVER DNA'S STRUCTURE FIRST. THE CHARGE REMAINS CONTROVERSIAL.
ROSALIND FRANKLIN
WATSON AND CRICK APPLIED THEIR RESEARCH-GUIDED THEORIES TO MODELS CONSTRUCTED OF CARDBOARD AND PASTE, REARRANGING MOLECULAR PLACEMENTS UNTIL THEY EVENTUALLY LOCKED TOGETHER IN LOGICAL ATOMIC BONDS. TOO MANY PIECES OF THE PUZZLE FIT FOR THEIR MODEL TO BE WRONG-- THE NOW-FAMILIAR DOUBLE-HELIX STRUCTURE OF DNA WAS REVEALED FOR THE FIRST TIME!
WATSON AND CRICK, ALONG WITH WILKINS, WERE AWARDED THE NOBEL PRIZE FOR MEDICINE IN 1962.
FRANKLIN, WHO HAD CONTRIBUTED MUCH TO THE DISCOVERY, WAS NOT ELIGIBLE FOR THE DISTINCTION. SHE DIED IN 1958 OF CANCER, A GENETICALLY RELATED DISEASE THAT SOMEDAY MAY BE BANISHED AS OUR UNDERSTANDING OF DNA GROWS.

THAT PRODUCTION IS CALLED THE TRANSLATION PROCESS. IT BEGINS WHEN A SMALL RIBOSOME SUBUNIT LATCHES ONTO ANY STRAND OF mRNA THAT HAS ENTERED THE CYTOPLASM.
SMALL RIBOSOME SUBUNIT
mRNA
THE mRNA IS CONVEYED THROUGH THE SMALL SUBUNIT UNTIL THE START CODON -- WHICH IS ALWAYS THE TRIPLET "AUG" -- IS LOCATED. WHEN THE SMALL SUBUNIT READS AUG, IT SENDS A SIGNAL CALLING THE APPROPRIATE MOLECULE OF TRANSFER RNA.
START CODON
AUGGAAUCA
tRNA MOLECULES ARE THE DELIVERY SYSTEMS OF THE TRANSLATION PROCESS -- THE TRUCKERS, IF YOU WILL. THEY HAUL THE APPROPRIATE AMINO ACID TO THE RIBOSOME, AS SPECIFIED BY THE CODON BEING READ.
SER
ARG
LYS
tRNA IS VERY SPECIALIZED, HAVING CONTORTED INTO A UNIQUE SHAPE NECESSARY TO ITS FUNCTION -- THANKS TO THE BONDING ABILITIES OF URACIL. EACH tRNA MOLECULE CARRIES A TRIPLET OF BASES CALLED AN ANTI-CODON, WHICH IS COMPLEMENTARY TO ITS SPECIFIC CODON ON mRNA.
mRNA CODON
AUGG
tRNA ANTI-CODON
UAC
A SPECIALIZED ENZYME READS tRNA'S ANTI-CODON, INFERS ITS CODON FROM THAT, AND ATTACHES ITS FREIGHT -- THE SPECIFIED AMINO ACID -- TO tRNA'S ACCEPTOR ARM.
UAC
AMINO ACID
SO, WHEN THE RIBOSOME READS AUG -- "START!" -- OFF THE mRNA, IT CALLS FOR A tRNA MOLECULE CARRYING THE AMINO ACID fMET TO LINK WITH THAT CODON.
AUGGAAUCAA
UAC
fMET

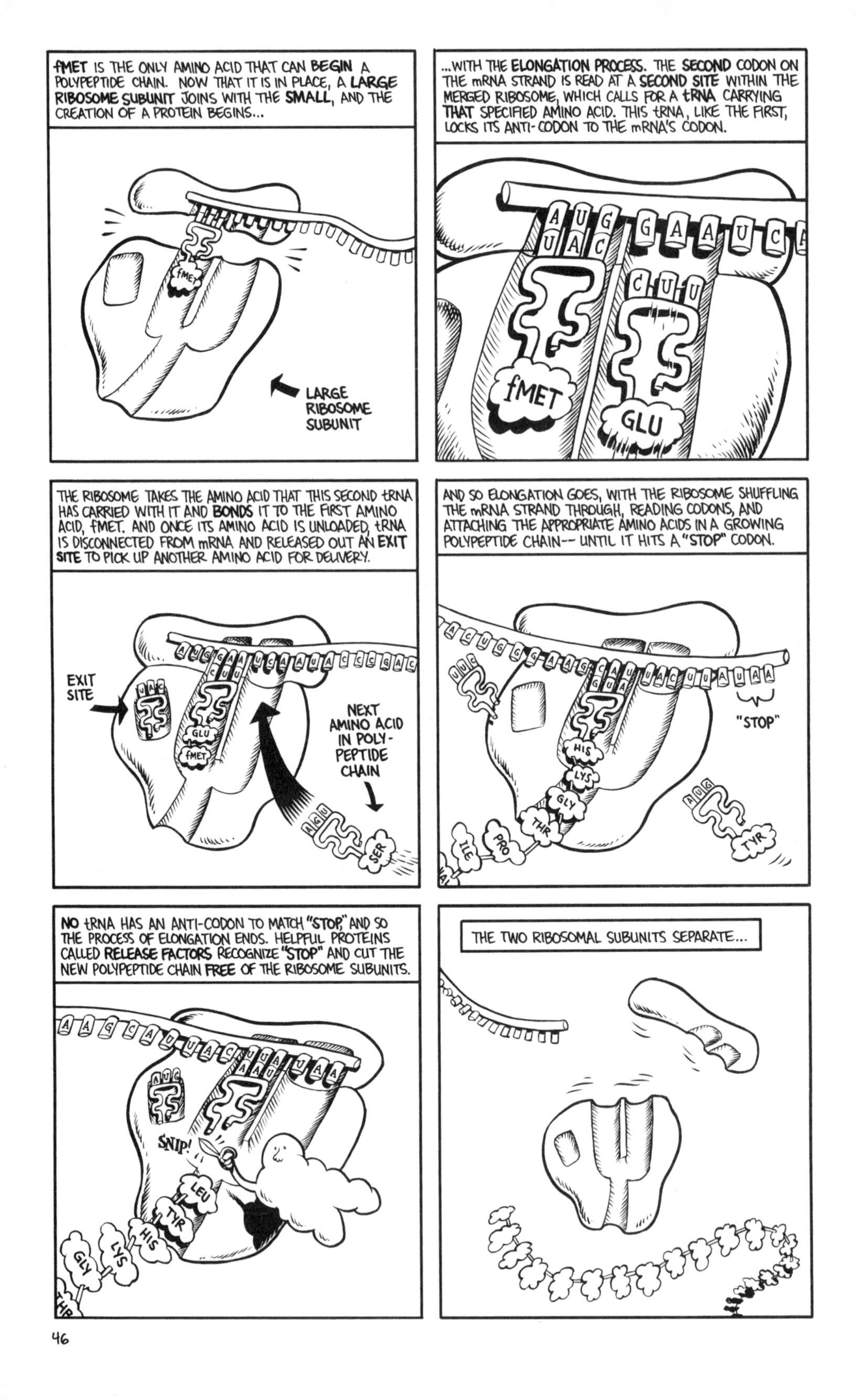
fMET IS THE ONLY AMINO ACID THAT CAN BEGIN A POLYPEPTIDE CHAIN. NOW THAT IT IS IN PLACE, A LARGE RIBOSOME SUBUNIT JOINS WITH THE SMALL, AND THE CREATION OF A PROTEIN BEGINS...
fMET
LARGE RIBOSOME SUBUNIT
...WITH THE ELONGATION PROCESS. THE SECOND CODON ON THE mRNA STRAND IS READ AT A SECOND SITE WITHIN THE MERGED RIBOSOME, WHICH CALLS FOR A tRNA CARRYING THAT SPECIFIED AMINO ACID. THIS tRNA, LIKE THE FIRST, LOCKS ITS ANTI-CODON TO THE mRNA'S CODON.
A U G G A A U C A
U A C
C U U
fMET
GLU
THE RIBOSOME TAKES THE AMINO ACID THAT THIS SECOND tRNA HAS CARRIED WITH IT AND BONDS IT TO THE FIRST AMINO ACID, fMET. AND ONCE ITS AMINO ACID IS UNLOADED, tRNA IS DISCONNECTED FROM mRNA AND RELEASED OUT AN EXIT SITE TO PICK UP ANOTHER AMINO ACID FOR DELIVERY.
EXIT SITE
GLU
fMET
NEXT AMINO ACID IN POLYPEPTIDE CHAIN
SER
AND SO ELONGATION GOES, WITH THE RIBOSOME SHUFFLING THE mRNA STRAND THROUGH, READING CODONS, AND ATTACHING THE APPROPRIATE AMINO ACIDS IN A GROWING POLYPEPTIDE CHAIN-- UNTIL IT HITS A "STOP" CODON.
"STOP"
HIS
LYS
GLY
THR
PRO
ILE
TYR
NO tRNA HAS AN ANTI-CODON TO MATCH "STOP," AND SO THE PROCESS OF ELONGATION ENDS. HELPFUL PROTEINS CALLED RELEASE FACTORS RECOGNIZE "STOP" AND CUT THE NEW POLYPEPTIDE CHAIN FREE OF THE RIBOSOME SUBUNITS.
SNIP!
LEU
TYR
HIS
LYS
GLY
THR
THE TWO RIBOSOMAL SUBUNITS SEPARATE...

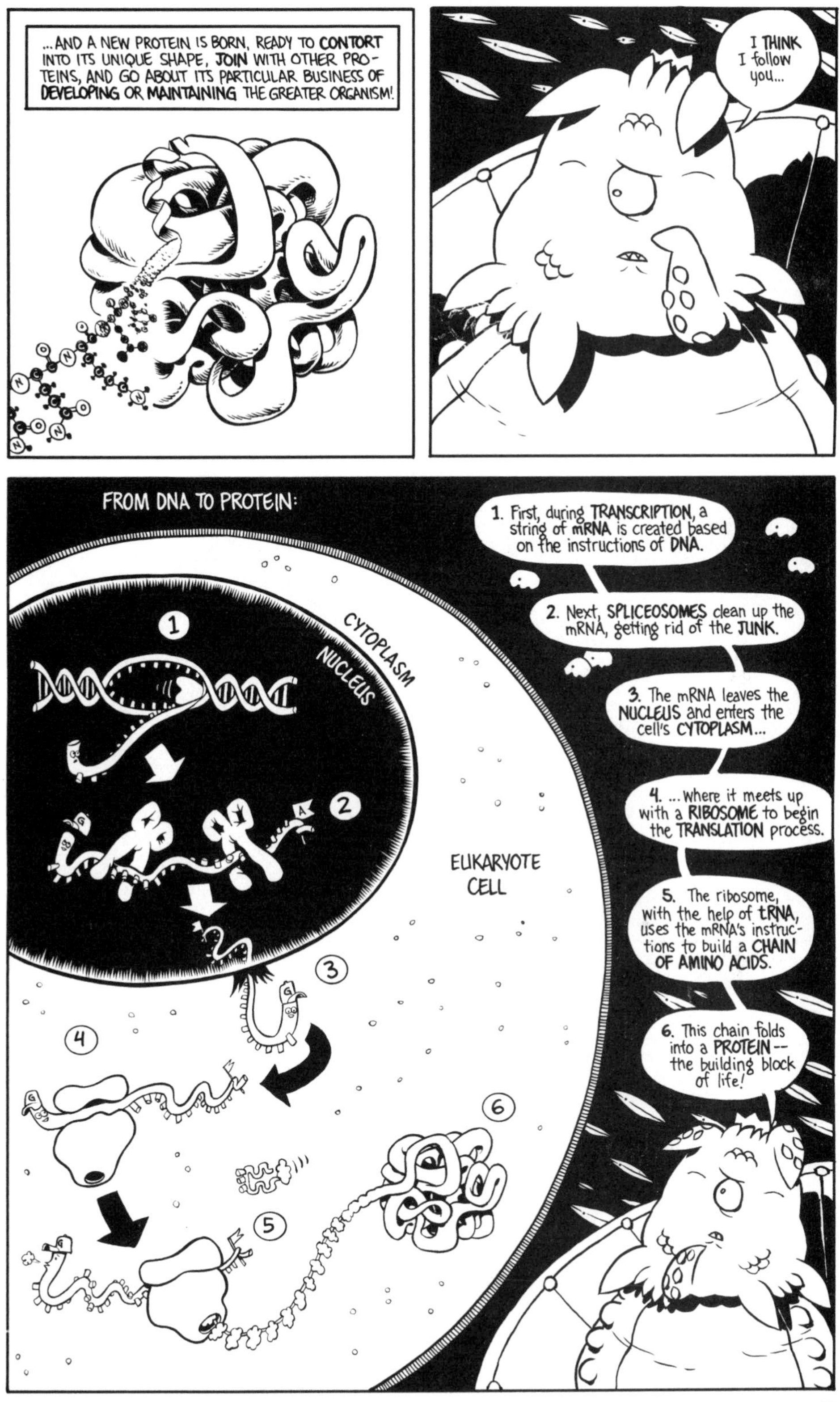
...AND A NEW PROTEIN IS BORN, READY TO CONTORT INTO ITS UNIQUE SHAPE, JOIN WITH OTHER PROTEINS, AND GO ABOUT ITS PARTICULAR BUSINESS OF DEVELOPING OR MAINTAINING THE GREATER ORGANISM!
I THINK I follow you...
FROM DNA TO PROTEIN:
CYTOPLASM
NUCLEUS
EUKARYOTE CELL
1. First, during TRANSCRIPTION, a string of mRNA is created based on the instructions of DNA.
2. Next, SPLICEOSOMES clean up the mRNA, getting rid of the JUNK.
3. The mRNA leaves the NUCLEUS and enters the cell's CYTOPLASM...
4. ...where it meets up with a RIBOSOME to begin the TRANSLATION process.
5. The ribosome, with the help of tRNA, uses the mRNA's instructions to build a CHAIN OF AMINO ACIDS.
6. This chain folds into a PROTEIN -- the building block of life!

Please tell me that's all there is to it!
ALMOST, SIRE!
RIBOSOMES AND tRNA, MEANWHILE, ARE FREE TO LOOK FOR OTHER mRNA STRANDS AND CONTINUE THE PROCESS OF BUILDING MORE POLYPEPTIDES. THEY ARE BOTH VERY RECYCLABLE.
OKAY, WHO'S NEXT? WHO'S NEXT?!
LYS
ARG
fMET
TYR
mRNA, HOWEVER, IS MADE TO DIE YOUNG. IT DELIVERS DNA'S INSTRUCTIONS TO THE DEGREE REQUIRED, AND THEN DISINTEGRATES, A GOOD SOLDIER TO THE END.
IT WAS FUN WHILE IT LASTED...
mRNA
AND THAT -- WHEW! -- CONCLUDES THE MOLECULAR PORTION OF MY REPORT.
...SIRE?...
mRNA
...YOUR ROYAL PERCEPTIVENESS?
Mind... spinning...
... At least THREE of my five brains must be overheated...
I KNOW, SIRE. I KNOW.
BUT, REST ASSURED, I'VE FRONT-LOADED MY REPORT WITH THE MOST TECHNICALLY CHALLENGING INFORMATION.
THE FOUNDATIONS OF THE LIFE PROCESS ARE DIFFICULT TO GRASP, INDEED...
... BUT A BASIC UNDERSTANDING OF DNA AND RNA AND WHAT THEY DO IS NECESSARY TO FULLY APPRECIATE THE REALLY GOOD STUFF.
I PROMISE YOU, THINGS GET A BIT LESS ARCANE FROM HERE ON, AS WE BEGIN TO LOOK AT...
...SEX!

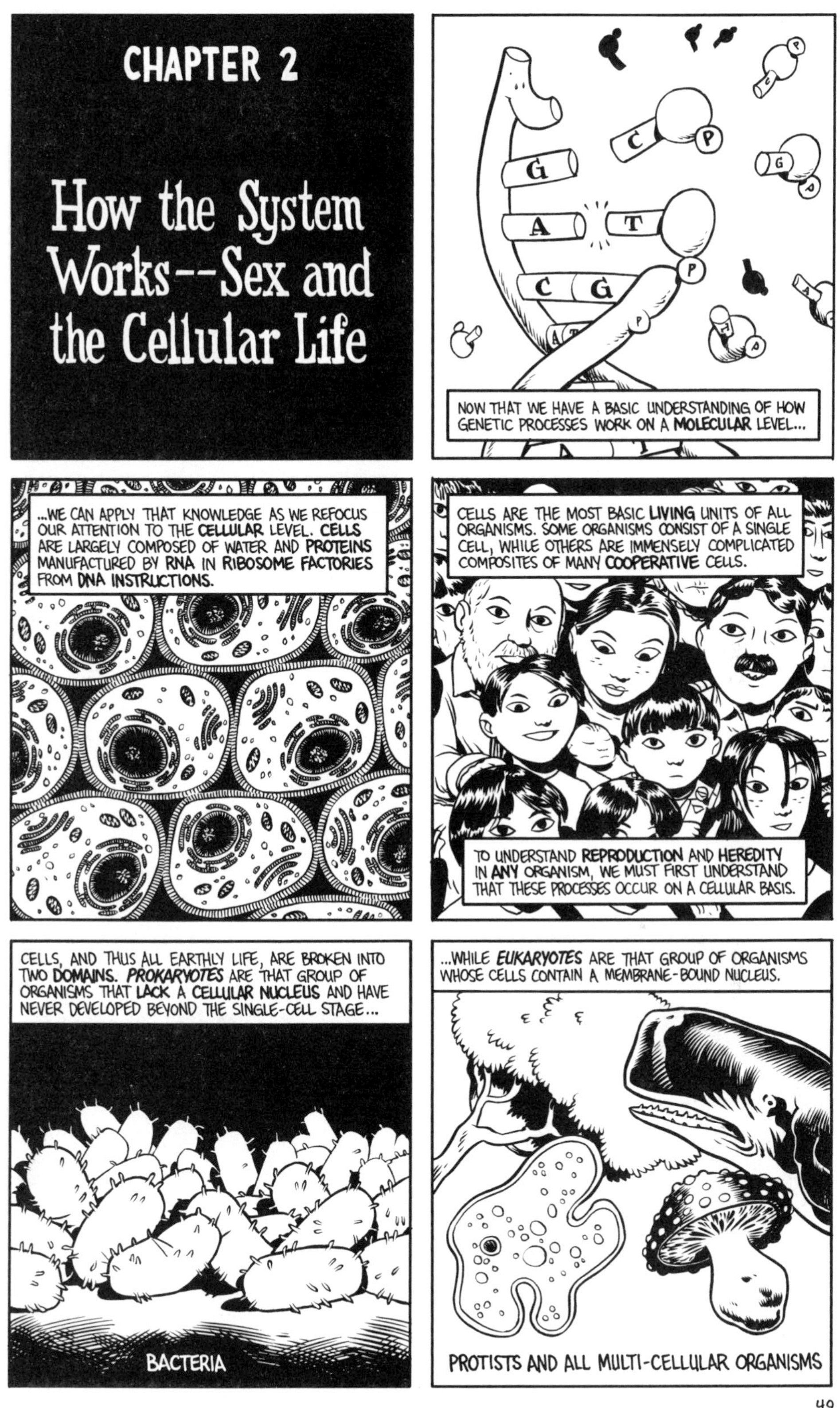
CHAPTER 2
How the System Works--Sex and the Cellular Life
G
A
T
C
G
C
P
NOW THAT WE HAVE A BASIC UNDERSTANDING OF HOW GENETIC PROCESSES WORK ON A MOLECULAR LEVEL...
...WE CAN APPLY THAT KNOWLEDGE AS WE REFOCUS OUR ATTENTION TO THE CELLULAR LEVEL. CELLS ARE LARGELY COMPOSED OF WATER AND PROTEINS MANUFACTURED BY RNA IN RIBOSOME FACTORIES FROM DNA INSTRUCTIONS.
CELLS ARE THE MOST BASIC LIVING UNITS OF ALL ORGANISMS. SOME ORGANISMS CONSIST OF A SINGLE CELL, WHILE OTHERS ARE IMMENSELY COMPLICATED COMPOSITES OF MANY COOPERATIVE CELLS.
TO UNDERSTAND REPRODUCTION AND HEREDITY IN ANY ORGANISM, WE MUST FIRST UNDERSTAND THAT THESE PROCESSES OCCUR ON A CELLULAR BASIS.
CELLS, AND THUS ALL EARTHLY LIFE, ARE BROKEN INTO TWO DOMAINS. PROKARYOTES ARE THAT GROUP OF ORGANISMS THAT LACK A CELLULAR NUCLEUS AND HAVE NEVER DEVELOPED BEYOND THE SINGLE-CELL STAGE...
BACTERIA
...WHILE EUKARYOTES ARE THAT GROUP OF ORGANISMS WHOSE CELLS CONTAIN A MEMBRANE-BOUND NUCLEUS.
PROTISTS AND ALL MULTI-CELLULAR ORGANISMS

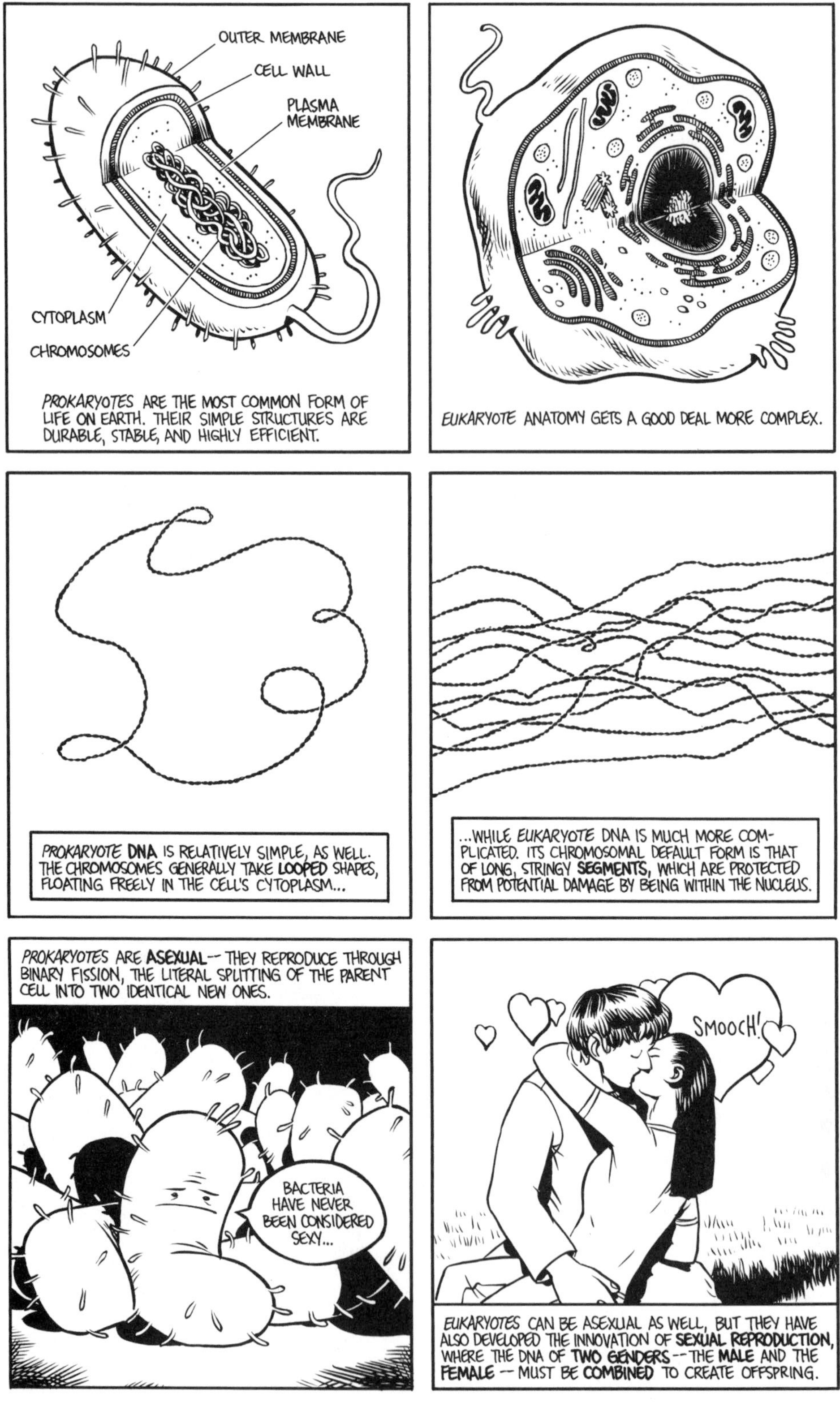
OUTER MEMBRANE
CELL WALL
PLASMA MEMBRANE
CYTOPLASM
CHROMOSOMES
PROKARYOTES ARE THE MOST COMMON FORM OF LIFE ON EARTH. THEIR SIMPLE STRUCTURES ARE DURABLE, STABLE, AND HIGHLY EFFICIENT.
EUKARYOTE ANATOMY GETS A GOOD DEAL MORE COMPLEX.
PROKARYOTE DNA IS RELATIVELY SIMPLE, AS WELL. THE CHROMOSOMES GENERALLY TAKE LOOPED SHAPES, FLOATING FREELY IN THE CELL'S CYTOPLASM...
...WHILE EUKARYOTE DNA IS MUCH MORE COMPLICATED. ITS CHROMOSOMAL DEFAULT FORM IS THAT OF LONG, STRINGY SEGMENTS, WHICH ARE PROTECTED FROM POTENTIAL DAMAGE BY BEING WITHIN THE NUCLEUS.
PROKARYOTES ARE ASEXUAL-- THEY REPRODUCE THROUGH BINARY FISSION, THE LITERAL SPLITTING OF THE PARENT CELL INTO TWO IDENTICAL NEW ONES.
BACTERIA HAVE NEVER BEEN CONSIDERED SEXY...
SMOOCH!
EUKARYOTES CAN BE ASEXUAL AS WELL, BUT THEY HAVE ALSO DEVELOPED THE INNOVATION OF SEXUAL REPRODUCTION, WHERE THE DNA OF TWO GENDERS--THE MALE AND THE FEMALE -- MUST BE COMBINED TO CREATE OFFSPRING.

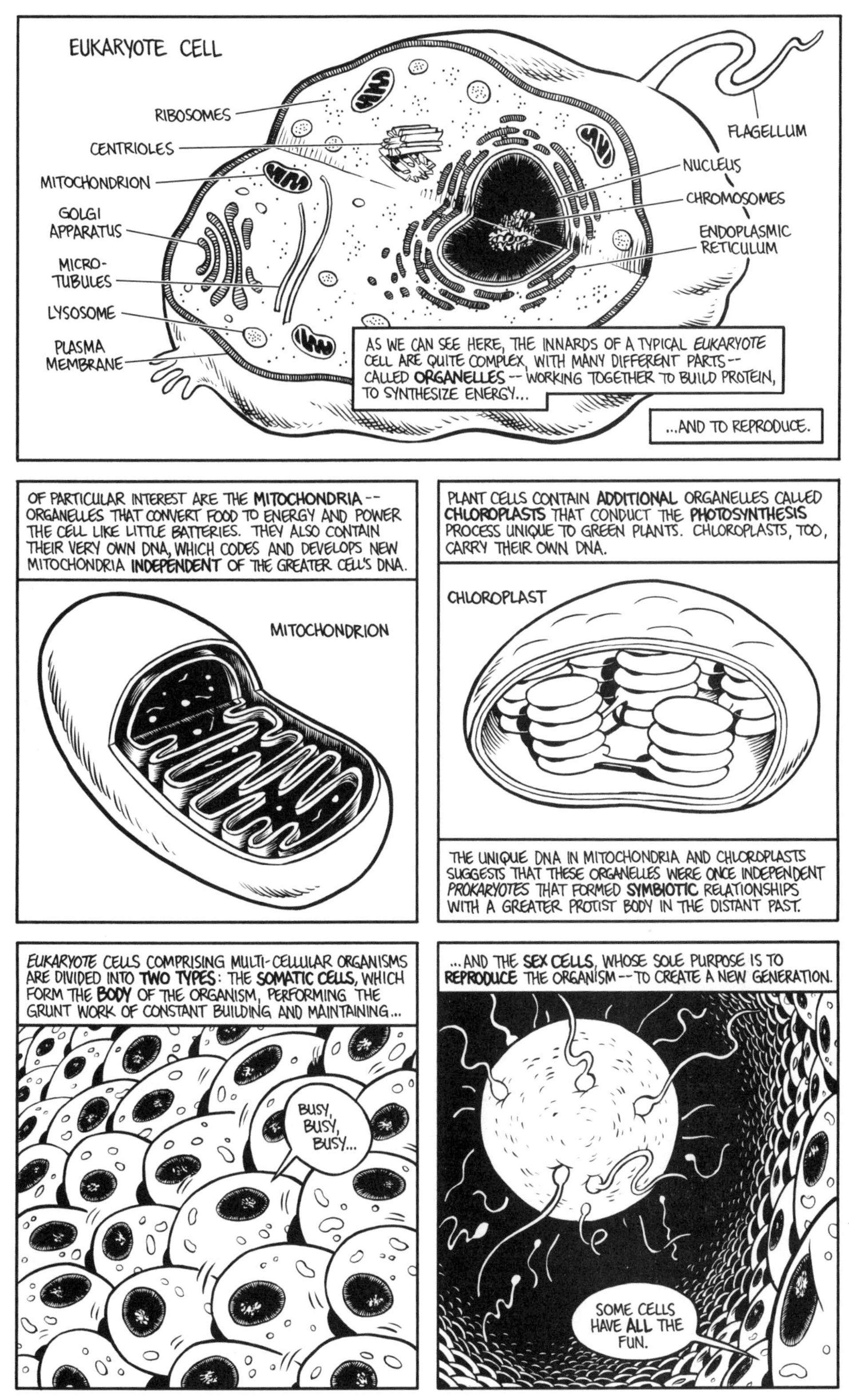
EUKARYOTE CELL
RIBOSOMES
CENTRIOLES
MITOCHONDRION
GOLGI APPARATUS
MICRO-TUBULES
LYSOSOME
PLASMA MEMBRANE
FLAGELLUM
NUCLEUS
CHROMOSOMES
ENDOPLASMIC RETICULUM
AS WE CAN SEE HERE, THE INNARDS OF A TYPICAL *EUKARYOTE* CELL ARE QUITE COMPLEX, WITH MANY DIFFERENT PARTS -- CALLED **ORGANELLES** -- WORKING TOGETHER TO BUILD PROTEIN, TO SYNTHESIZE ENERGY...
...AND TO REPRODUCE.
OF PARTICULAR INTEREST ARE THE **MITOCHONDRIA** -- ORGANELLES THAT CONVERT FOOD TO ENERGY AND POWER THE CELL LIKE LITTLE BATTERIES. THEY ALSO CONTAIN THEIR VERY OWN DNA, WHICH CODES AND DEVELOPS NEW MITOCHONDRIA **INDEPENDENT** OF THE GREATER CELL'S DNA.
MITOCHONDRION
PLANT CELLS CONTAIN **ADDITIONAL** ORGANELLES CALLED **CHLOROPLASTS** THAT CONDUCT THE **PHOTOSYNTHESIS** PROCESS UNIQUE TO GREEN PLANTS. CHLOROPLASTS, TOO, CARRY THEIR OWN DNA.
CHLOROPLAST
THE UNIQUE DNA IN MITOCHONDRIA AND CHLOROPLASTS SUGGESTS THAT THESE ORGANELLES WERE ONCE INDEPENDENT *PROKARYOTES* THAT FORMED **SYMBIOTIC** RELATIONSHIPS WITH A GREATER PROTIST BODY IN THE DISTANT PAST.
EUKARYOTE CELLS COMPRISING MULTI-CELLULAR ORGANISMS ARE DIVIDED INTO **TWO TYPES**: THE **SOMATIC CELLS**, WHICH FORM THE **BODY** OF THE ORGANISM, PERFORMING THE GRUNT WORK OF CONSTANT BUILDING AND MAINTAINING...
BUSY, BUSY, BUSY...
...AND THE **SEX CELLS**, WHOSE SOLE PURPOSE IS TO **REPRODUCE** THE ORGANISM -- TO CREATE A NEW GENERATION.
SOME CELLS HAVE **ALL** THE FUN.

BECAUSE SOMATIC CELLS MUST PERFORM ALL THE MANY DIVERSE FUNCTIONS NECESSARY TO GROW, MAINTAIN, AND REPAIR AN ORGANISM, THEY CARRY THE ORGANISM'S **COMPLETE** GENOME WITH THEM, EXPRESSING THE GENES REQUIRED FOR THEIR SPECIALIZATION.
FAT
LEAF
BONE
THEY REPRODUCE THEMSELVES BY SIMPLE **MITOSIS**--THE PROCESS THAT DUPLICATES **EXACTLY** THE CELL'S COMPLETE SET OF CHROMOSOMES AND PASSES IT UNALTERED FROM ONE GENERATION TO THE NEXT.
BONE
SAME BONE
SEX CELLS, ON THE OTHER HAND, HAVE A SPECIAL **PROBLEM** TO OVERCOME, AS THE SEXUAL REPRODUCTION STRATEGY REQUIRES THAT THE GENETIC MATERIAL OF **TWO** INDIVIDUAL CELLS BE COMBINED.
PROBLEM? WHAT PROBLEM?
SPECIFICALLY, SEX CELLS MUST FIND A WAY TO **HALVE** THEIR FULL COMPLEMENT OF DNA, SO THAT WHEN COMBINED, A CORRECT, PRECISE COMPLEMENT OF DNA IS FORMED FROM THE TWO CELLS.
THIS PROCESS OF HALVING THE SEX CELL'S DNA IS CALLED **MEIOSIS**.
ALL IT TAKES IS **MILLIONS** OF YEARS OF EXPERIENCE.
BEFORE WE GO ANY FURTHER, WE NEED TO TAKE A CLOSER LOOK AT **CHROMOSOMES**--THOSE PHYSICALLY ORGANIZED FORMS OF DNA THAT CARRY THE CELL'S GENES.
EVERY EUKARYOTE CARRIES ITS GENETIC HERITAGE IN A NUMBER OF CHROMOSOMES THAT IS **STANDARD** TO ITS SPECIES.
MAIZE CARRIES 20 CHROMOSOMES...

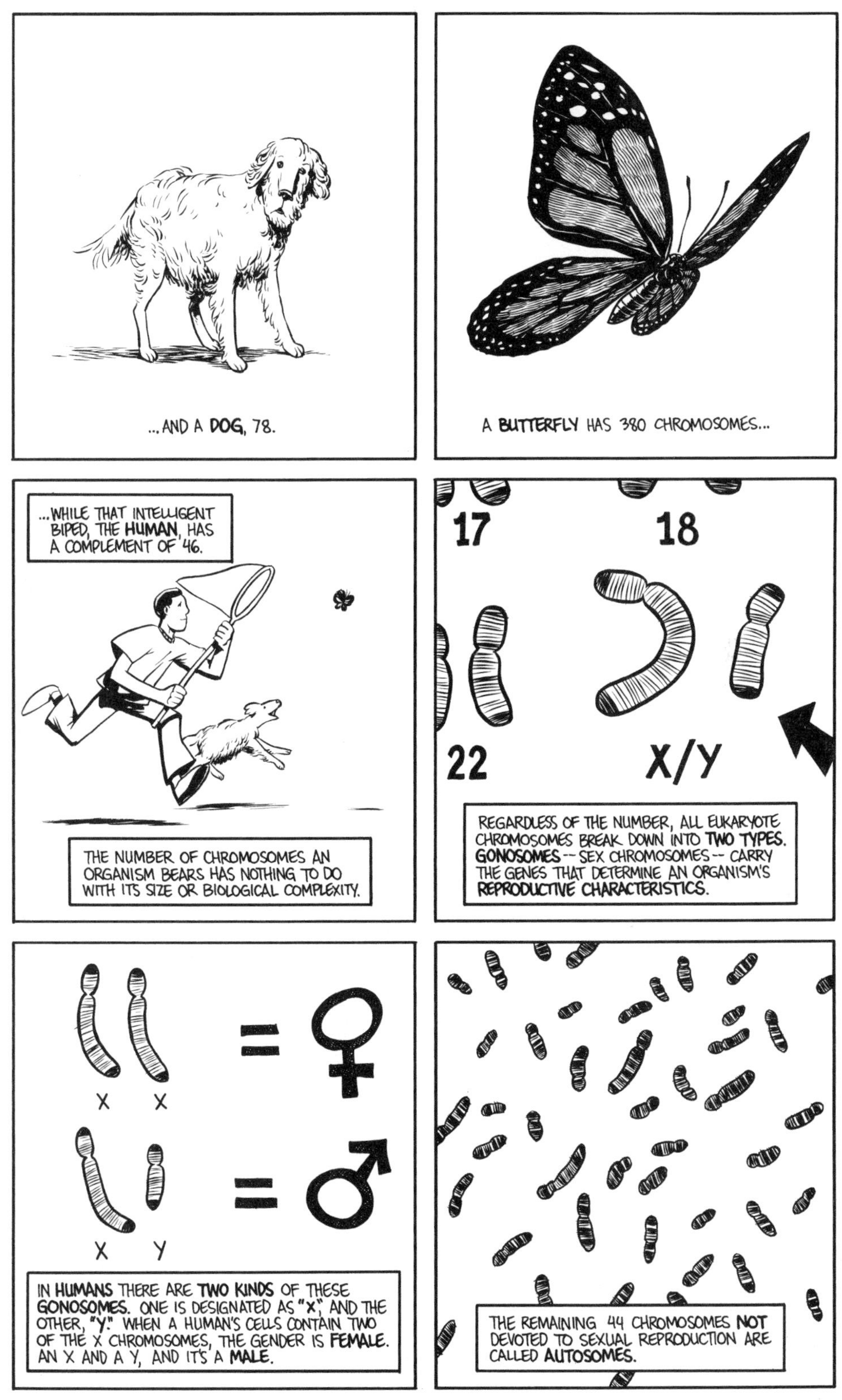
...AND A **DOG**, 78.
A **BUTTERFLY** HAS 380 CHROMOSOMES...
...WHILE THAT INTELLIGENT BIPED, THE **HUMAN**, HAS A COMPLEMENT OF 46.
THE NUMBER OF CHROMOSOMES AN ORGANISM BEARS HAS NOTHING TO DO WITH ITS SIZE OR BIOLOGICAL COMPLEXITY.
17
18
22
X/Y
REGARDLESS OF THE NUMBER, ALL EUKARYOTE CHROMOSOMES BREAK DOWN INTO **TWO TYPES**. **GONOSOMES** -- SEX CHROMOSOMES -- CARRY THE GENES THAT DETERMINE AN ORGANISM'S **REPRODUCTIVE CHARACTERISTICS**.
X X
X Y
IN **HUMANS** THERE ARE **TWO KINDS** OF THESE **GONOSOMES**. ONE IS DESIGNATED AS "**X**," AND THE OTHER, "**Y**." WHEN A HUMAN'S CELLS CONTAIN TWO OF THE X CHROMOSOMES, THE GENDER IS **FEMALE**. AN X AND A Y, AND IT'S A **MALE**.
THE REMAINING 44 CHROMOSOMES **NOT** DEVOTED TO SEXUAL REPRODUCTION ARE CALLED **AUTOSOMES**.

1
2
3
4
5
6
7
8
9
10
11
12
13
14
15
16
17
18
19
20
21
22
THE 44 AUTOSOMES ARE ACTUALLY **TWIN SETS OF 22 CHROMOSOMES.** IN OTHER WORDS, HUMANS HAVE 22 UNIQUE NON-SEX CHROMOSOMES, EACH OF WHICH HAS A **DUPLICATE.**
EACH OF THE CHROMOSOMES IN A PAIRED SET OF DUPLICATES IS CALLED A **HOMOLOG,** AND THE TWO CHROMOSOMES' RELATIONSHIP IS DESCRIBED AS **HOMOLOGOUS.**
MY **HOMEY!**
THIS BIOLOGICAL STRATEGY OF CARRYING TWO COPIES OF CHROMOSOMES IS A COMMON ONE AMONG EARTHLY *EUKARYOTES*--AND THOSE EMPLOYING IT ARE TERMED **DIPLOID.**
BUT IT IS NOT THE ONLY GAME IN TOWN...
THE DIPLOID CLUB
...THERE ARE FERNS, FOR INSTANCE, THAT CARRY AS MANY AS 630 PAIRS OF CHROMOSOMES! SOME BEES, ANTS, AND WASPS, ON THE OTHER EXTREME, ARE **HAPLOID** ORGANISMS--MEANING THEY CARRY ONLY **ONE** SET OF CHROMOSOMES.
AND WE DO JUST FINE, THANK YOU.
WHICH BRINGS US BACK AGAIN TO OUR **HUMAN** EXAMPLE--WHILE HUMAN AUTOSOME CHROMOSOMES ARE DIPLOID, THE **GONOSOMES**--THE TWO SEX CHROMOSOMES--ARE **HAPLOID.**
I AM A UNIQUE INDIVIDUAL.
AS AM I. ONE OF A KIND.
THERE IS A VERY IMPORTANT REASON FOR **HAPLOIDY** IN SEX CELLS, AS WE'LL COME TO SEE, BUT **FIRST...**

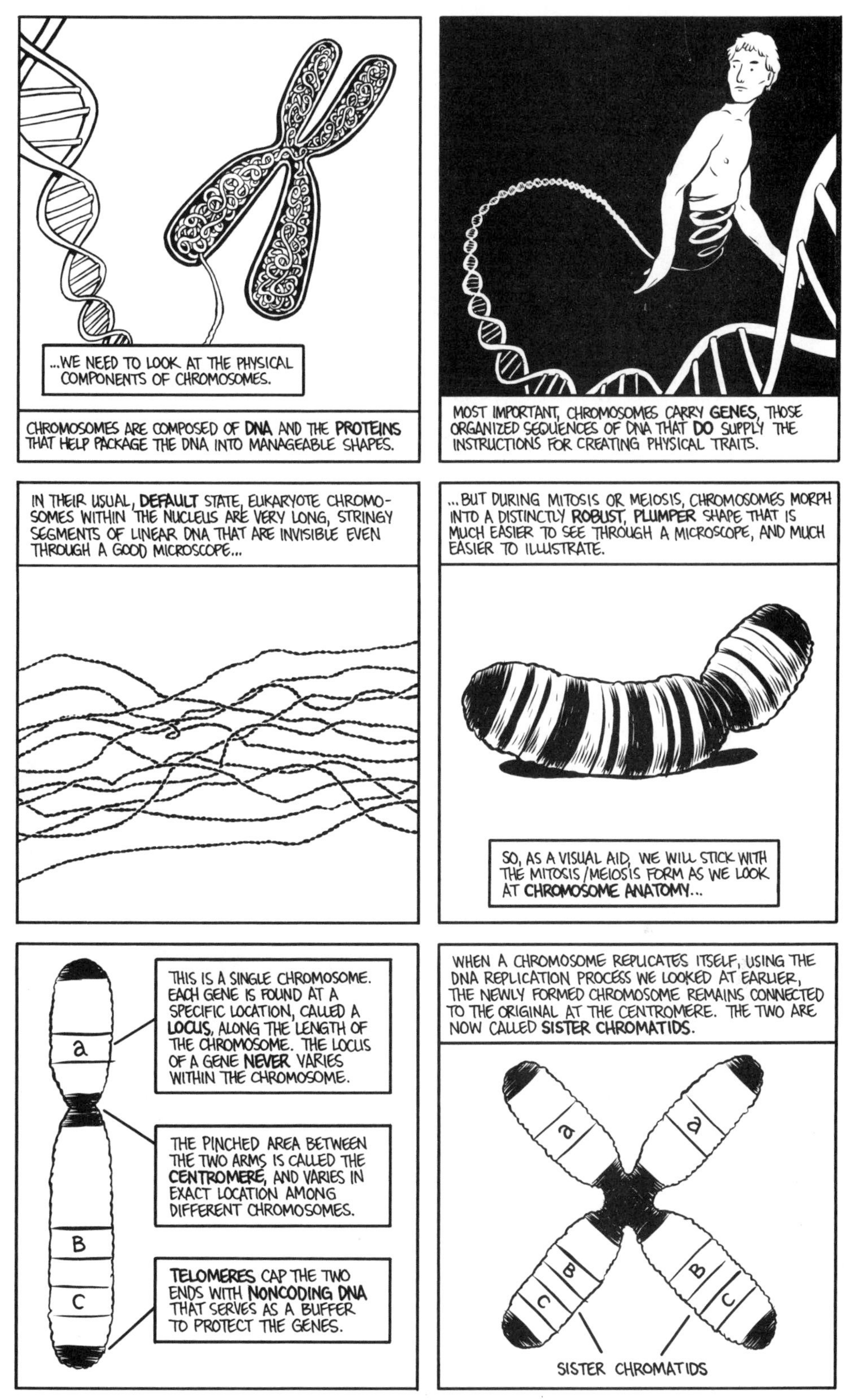

...WE NEED TO LOOK AT THE PHYSICAL COMPONENTS OF CHROMOSOMES.
CHROMOSOMES ARE COMPOSED OF **DNA** AND THE **PROTEINS** THAT HELP PACKAGE THE DNA INTO MANAGEABLE SHAPES.
MOST IMPORTANT, CHROMOSOMES CARRY **GENES**, THOSE ORGANIZED SEQUENCES OF DNA THAT **DO** SUPPLY THE INSTRUCTIONS FOR CREATING PHYSICAL TRAITS.
IN THEIR USUAL, **DEFAULT** STATE, EUKARYOTE CHROMOSOMES WITHIN THE NUCLEUS ARE VERY LONG, STRINGY SEGMENTS OF LINEAR DNA THAT ARE INVISIBLE EVEN THROUGH A GOOD MICROSCOPE...
...BUT DURING MITOSIS OR MEIOSIS, CHROMOSOMES MORPH INTO A DISTINCTLY **ROBUST**, **PLUMPER** SHAPE THAT IS MUCH EASIER TO SEE THROUGH A MICROSCOPE, AND MUCH EASIER TO ILLUSTRATE.
SO, AS A VISUAL AID, WE WILL STICK WITH THE MITOSIS/MEIOSIS FORM AS WE LOOK AT **CHROMOSOME ANATOMY**...
a
B
C
THIS IS A SINGLE CHROMOSOME. EACH GENE IS FOUND AT A SPECIFIC LOCATION, CALLED A **LOCUS**, ALONG THE LENGTH OF THE CHROMOSOME. THE LOCUS OF A GENE **NEVER** VARIES WITHIN THE CHROMOSOME.
THE PINCHED AREA BETWEEN THE TWO ARMS IS CALLED THE **CENTROMERE**, AND VARIES IN EXACT LOCATION AMONG DIFFERENT CHROMOSOMES.
TELOMERES CAP THE TWO ENDS WITH **NONCODING DNA** THAT SERVES AS A BUFFER TO PROTECT THE GENES.
WHEN A CHROMOSOME REPLICATES ITSELF, USING THE DNA REPLICATION PROCESS WE LOOKED AT EARLIER, THE NEWLY FORMED CHROMOSOME REMAINS CONNECTED TO THE ORIGINAL AT THE CENTROMERE. THE TWO ARE NOW CALLED **SISTER CHROMATIDS**.
a
a
B
C
B
C
SISTER CHROMATIDS

NOW WE CAN GET INTO AN ASPECT OF CHROMOSOME ANATOMY THAT IS OF CENTRAL IMPORTANCE TO SEXUAL REPRODUCTION!
THESE TWO SISTER CHROMATIDS ARE HOMOLOGOUS. THEY ARE DIPLOID AUTOSOMES, MEANING THEY BEAR IDENTICAL COPIES OF ALL CHROMOSOMES NOT RELATED TO SEX. THE GENES THEY CARRY SIT AT THE SAME LOCI ON EITHER HOMOLOG.
BUT--THOSE SAME GENES ARE NOT NECESSARILY IDENTICAL! EACH HOMOLOG MAY HAVE VARIANTS OF THE SAME GENE. THESE ARE CALLED ALLELES, AND ARE INDICATED IN NOTATION BY THE USE OF THE UPPER- OR LOWERCASE OF THE SAME LETTER.
a
a
B
B
C
C
A
A
b
b
c
c
HOMOLOGS
A GENE THAT CODES FOR A PARTICULAR PHYSICAL TRAIT--LET'S SAY HAIR COLOR--CAN COME IN TWO OR MORE VARIANTS.
ONE ALLELE OF THAT HAIR-COLOR GENE MAY CODE FOR BROWN...
BB
...WHILE THE OTHER ALLELE PRODUCES BLOND HAIR.
IT IS THE ALLELES THAT ALLOW FOR THE GREAT VARIATION WE SEE WITHIN ORGANISMS THAT REPRODUCE SEXUALLY!
bb
NOW, UNDERSTAND THAT THIS EXAMPLE IS EXTREMELY SIMPLIFIED...
...MOST OF AN INDIVIDUAL ORGANISM'S PHYSICAL TRAITS ARE CREATED THROUGH THE INTERACTION OF A NUMBER OF ALLELES, AS WELL AS ENVIRONMENTAL FACTORS...
bb
...But I'm starting to get the picture.
All these chromosome complexities create variety within a particular species of organism.
WHEN APPLIED TO SEXUAL REPRODUCTION, YES--THEY DO!

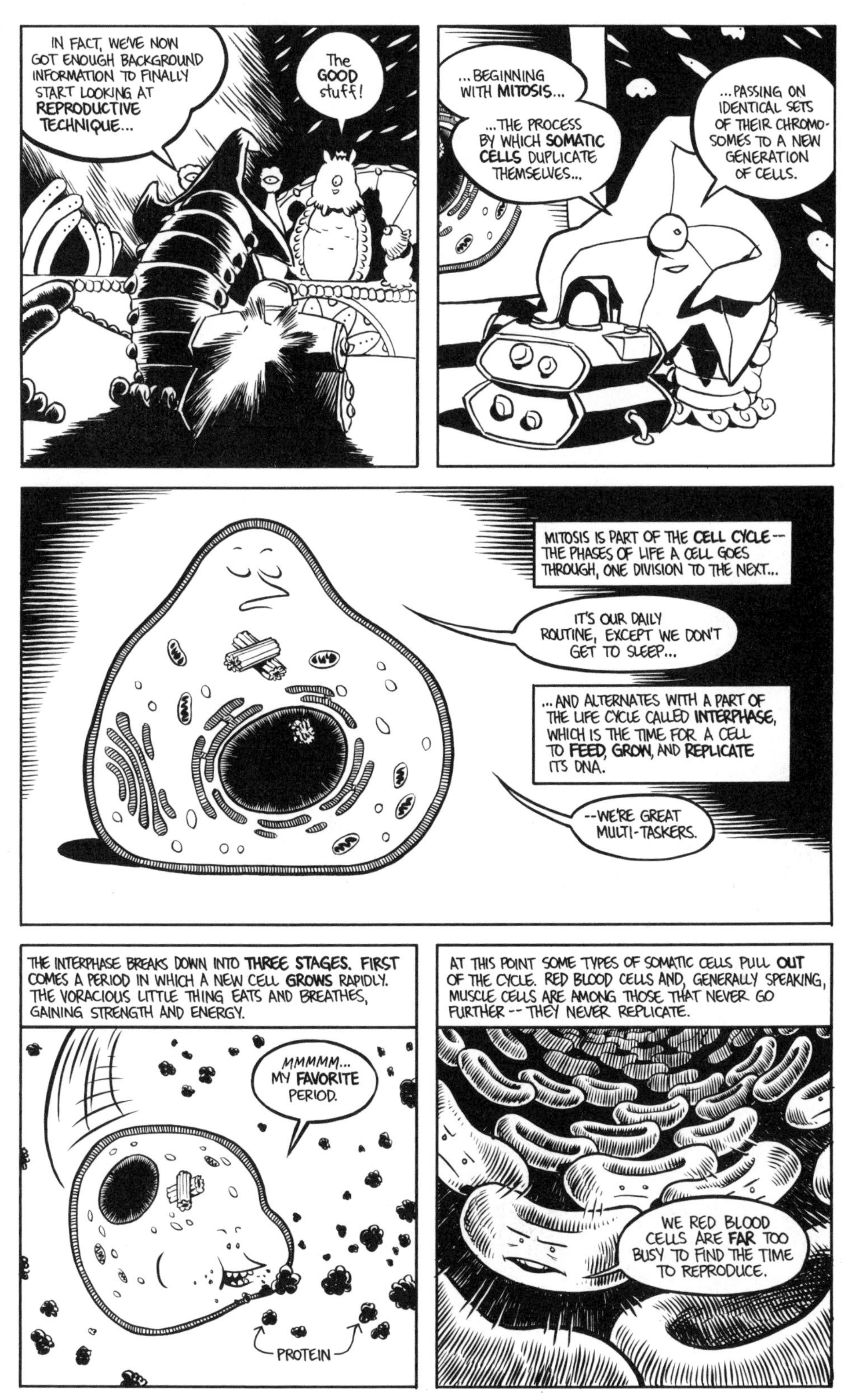
IN FACT, WE'VE NOW GOT ENOUGH BACKGROUND INFORMATION TO FINALLY START LOOKING AT REPRODUCTIVE TECHNIQUE...
The GOOD stuff!
...BEGINNING WITH MITOSIS...
...THE PROCESS BY WHICH SOMATIC CELLS DUPLICATE THEMSELVES...
...PASSING ON IDENTICAL SETS OF THEIR CHROMOSOMES TO A NEW GENERATION OF CELLS.
MITOSIS IS PART OF THE CELL CYCLE--THE PHASES OF LIFE A CELL GOES THROUGH, ONE DIVISION TO THE NEXT...
IT'S OUR DAILY ROUTINE, EXCEPT WE DON'T GET TO SLEEP...
...AND ALTERNATES WITH A PART OF THE LIFE CYCLE CALLED INTERPHASE, WHICH IS THE TIME FOR A CELL TO FEED, GROW, AND REPLICATE ITS DNA.
--WE'RE GREAT MULTI-TASKERS.
THE INTERPHASE BREAKS DOWN INTO THREE STAGES. FIRST COMES A PERIOD IN WHICH A NEW CELL GROWS RAPIDLY. THE VORACIOUS LITTLE THING EATS AND BREATHES, GAINING STRENGTH AND ENERGY.
MMMMM... MY FAVORITE PERIOD.
PROTEIN
AT THIS POINT SOME TYPES OF SOMATIC CELLS PULL OUT OF THE CYCLE. RED BLOOD CELLS AND, GENERALLY SPEAKING, MUSCLE CELLS ARE AMONG THOSE THAT NEVER GO FURTHER -- THEY NEVER REPLICATE.
WE RED BLOOD CELLS ARE FAR TOO BUSY TO FIND THE TIME TO REPRODUCE.

MOST CELLS, HOWEVER, **CONTINUE** THROUGH THE CYCLE TO THE **SECOND STAGE**, IN WHICH THE **REPLICATION** OF THE CELL'S **NUCLEAR DNA** TAKES PLACE. ALL THE CHROMOSOMES ARE DUPLICATED, BECOMING **SISTER CHROMATIDS** CONNECTED AT THEIR CENTROMERES.
AFTER DNA REPLICATION IS COMPLETE, THE CELL TAKES ANOTHER SHORT BREATHER, GROWING SOME MORE—THE **THIRD** STAGE OF METAPHASE—BEFORE FINALLY ENTERING...
WHEW!
IT TAKES A LOT OF ENERGY TO BE A GOOD CELL...
...**MITOSIS**...
...THE SEPARATION OF THE 46 CONJOINED SISTER CHROMATIDS INTO **TWO CELLS**.
THERE ARE FIVE **STAGES** TO THIS PROCESS.
AND, BY THE WAY, **PLANT** AND **ANIMAL** MITOSIS **DIFFER** A BIT, SO WE'RE GOING TO STICK WITH THE **ANIMAL** PROCESS HERE.
FIRST COMES **PROPHASE**, IN WHICH THE CHROMOSOMES CONTRACT AND CONDENSE INTO THEIR ROBUST, PLUMP, VISIBLE SHAPES.
PROPHASE
TO 19TH-CENTURY SCIENTISTS, THE FIRST OBSERVERS OF CHROMOSOMES, THIS CHANGE IN FORM SEEMED TO **POP** THESE MICROSCOPIC SQUIGGLES INTO EXISTENCE FROM OUT OF NOWHERE!
THERE WAS NOTHING THERE A MINUTE AGO!
AT THE SAME TIME THAT THE CHROMOSOMES COMPACT INTO THEIR NEW SHAPES, THE CELL NUCLEUS BEGINS TO **DISINTEGRATE**, ALLOWING THE CHROMOSOMES TO MIGRATE INTO OPEN **CYTOPLASM**.
FREE AT LAST!

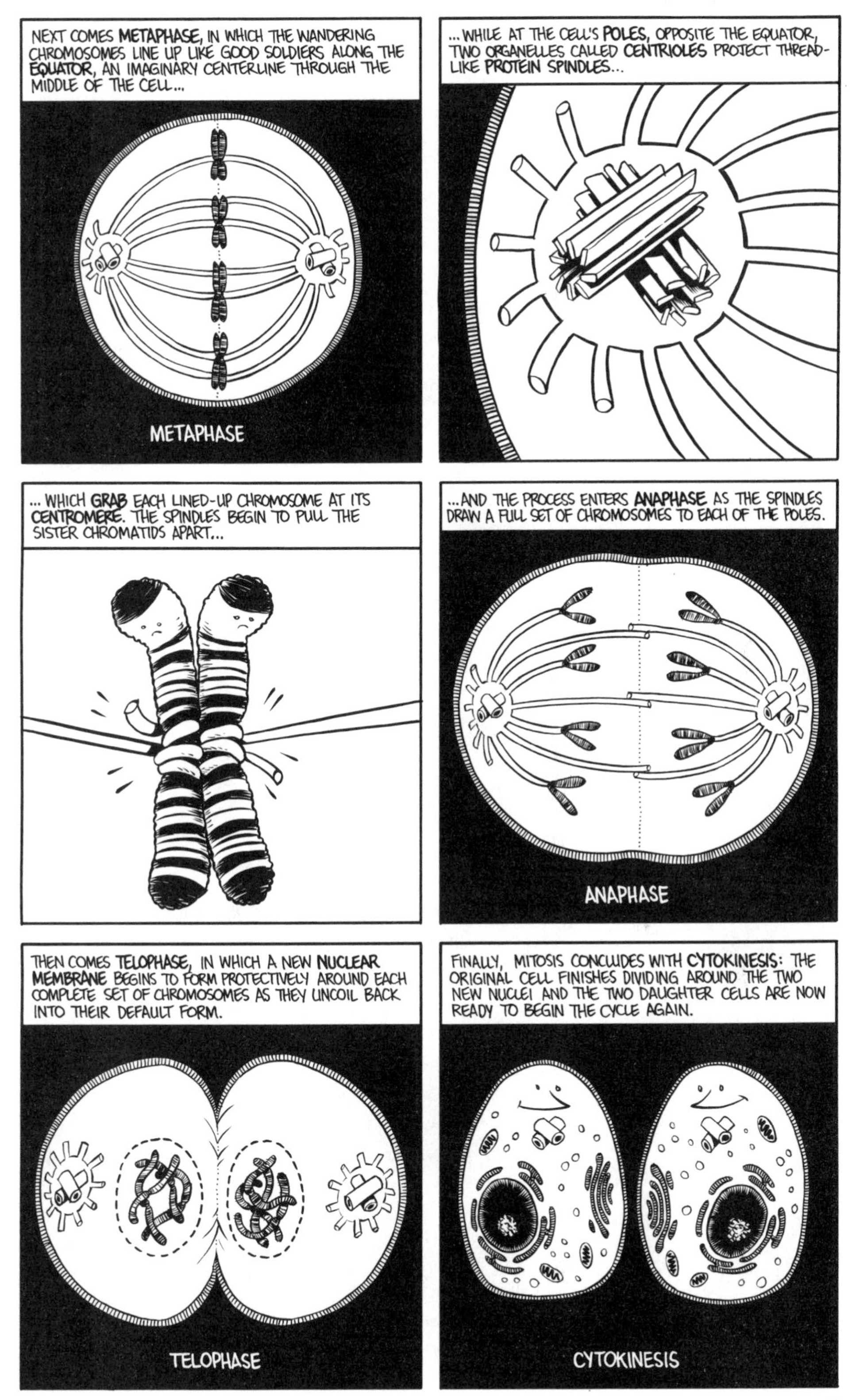
NEXT COMES METAPHASE, IN WHICH THE WANDERING CHROMOSOMES LINE UP LIKE GOOD SOLDIERS ALONG THE EQUATOR, AN IMAGINARY CENTERLINE THROUGH THE MIDDLE OF THE CELL...
METAPHASE
...WHILE AT THE CELL'S POLES, OPPOSITE THE EQUATOR, TWO ORGANELLES CALLED CENTRIOLES PROJECT THREAD-LIKE PROTEIN SPINDLES...
... WHICH GRAB EACH LINED-UP CHROMOSOME AT ITS CENTROMERE. THE SPINDLES BEGIN TO PULL THE SISTER CHROMATIDS APART...
...AND THE PROCESS ENTERS ANAPHASE AS THE SPINDLES DRAW A FULL SET OF CHROMOSOMES TO EACH OF THE POLES.
ANAPHASE
THEN COMES TELOPHASE, IN WHICH A NEW NUCLEAR MEMBRANE BEGINS TO FORM PROTECTIVELY AROUND EACH COMPLETE SET OF CHROMOSOMES AS THEY UNCOIL BACK INTO THEIR DEFAULT FORM.
TELOPHASE
FINALLY, MITOSIS CONCLUDES WITH CYTOKINESIS: THE ORIGINAL CELL FINISHES DIVIDING AROUND THE TWO NEW NUCLEI AND THE TWO DAUGHTER CELLS ARE NOW READY TO BEGIN THE CYCLE AGAIN.
CYTOKINESIS

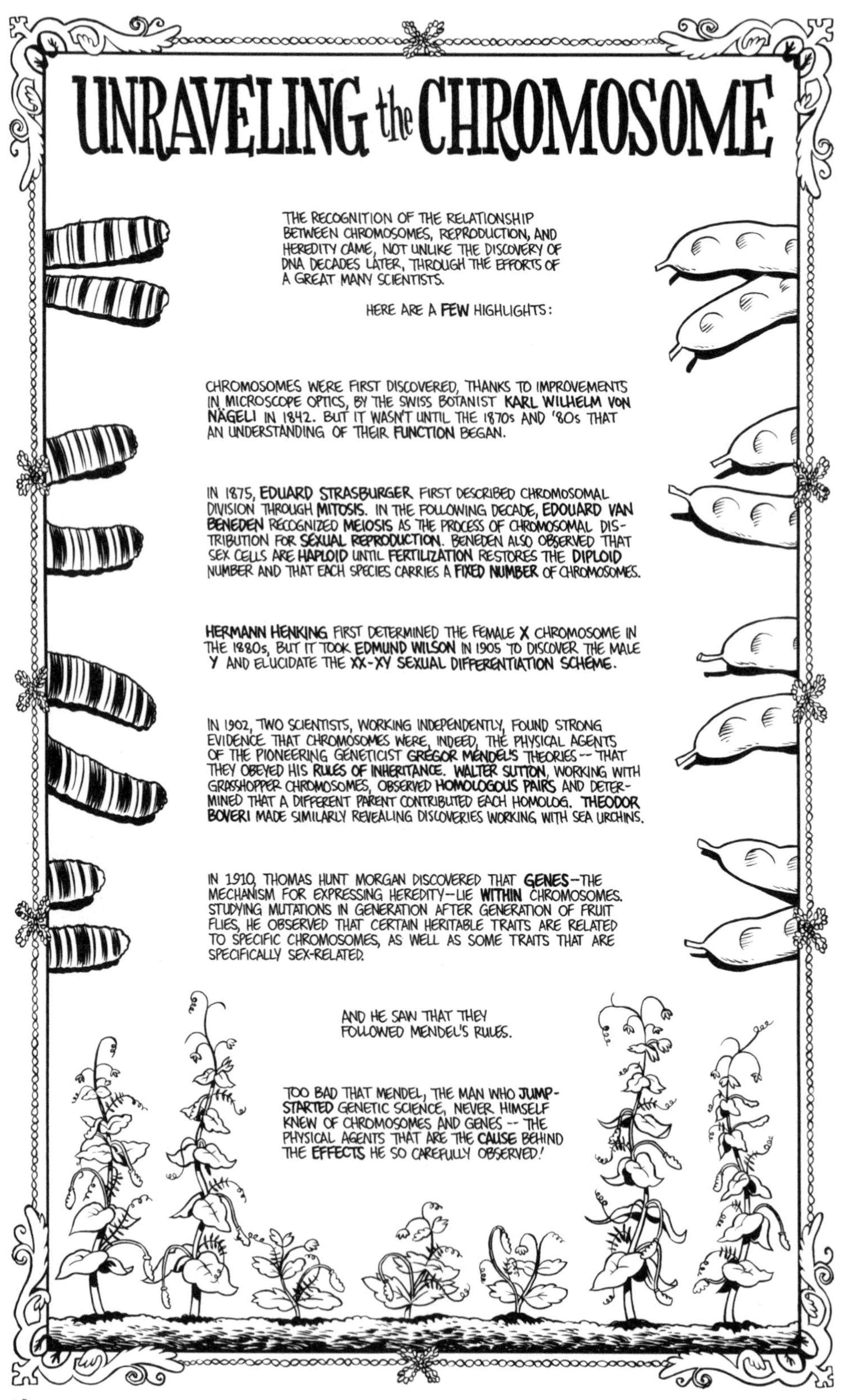
UNRAVELING the CHROMOSOME
THE RECOGNITION OF THE RELATIONSHIP BETWEEN CHROMOSOMES, REPRODUCTION, AND HEREDITY CAME, NOT UNLIKE THE DISCOVERY OF DNA DECADES LATER, THROUGH THE EFFORTS OF A GREAT MANY SCIENTISTS.
HERE ARE A FEW HIGHLIGHTS:
CHROMOSOMES WERE FIRST DISCOVERED, THANKS TO IMPROVEMENTS IN MICROSCOPE OPTICS, BY THE SWISS BOTANIST KARL WILHELM VON NÄGELI IN 1842. BUT IT WASN'T UNTIL THE 1870s AND '80s THAT AN UNDERSTANDING OF THEIR FUNCTION BEGAN.
IN 1875, EDUARD STRASBURGER FIRST DESCRIBED CHROMOSOMAL DIVISION THROUGH MITOSIS. IN THE FOLLOWING DECADE, EDOUARD VAN BENEDEN RECOGNIZED MEIOSIS AS THE PROCESS OF CHROMOSOMAL DISTRIBUTION FOR SEXUAL REPRODUCTION. BENEDEN ALSO OBSERVED THAT SEX CELLS ARE HAPLOID UNTIL FERTILIZATION RESTORES THE DIPLOID NUMBER AND THAT EACH SPECIES CARRIES A FIXED NUMBER OF CHROMOSOMES.
HERMANN HENKING FIRST DETERMINED THE FEMALE X CHROMOSOME IN THE 1880s, BUT IT TOOK EDMUND WILSON IN 1905 TO DISCOVER THE MALE Y AND ELUCIDATE THE XX-XY SEXUAL DIFFERENTIATION SCHEME.
IN 1902, TWO SCIENTISTS, WORKING INDEPENDENTLY, FOUND STRONG EVIDENCE THAT CHROMOSOMES WERE, INDEED, THE PHYSICAL AGENTS OF THE PIONEERING GENETICIST GREGOR MENDEL'S THEORIES -- THAT THEY OBEYED HIS RULES OF INHERITANCE. WALTER SUTTON, WORKING WITH GRASSHOPPER CHROMOSOMES, OBSERVED HOMOLOGOUS PAIRS AND DETERMINED THAT A DIFFERENT PARENT CONTRIBUTED EACH HOMOLOG. THEODOR BOVERI MADE SIMILARLY REVEALING DISCOVERIES WORKING WITH SEA URCHINS.
IN 1910, THOMAS HUNT MORGAN DISCOVERED THAT GENES--THE MECHANISM FOR EXPRESSING HEREDITY--LIE WITHIN CHROMOSOMES. STUDYING MUTATIONS IN GENERATION AFTER GENERATION OF FRUIT FLIES, HE OBSERVED THAT CERTAIN HERITABLE TRAITS ARE RELATED TO SPECIFIC CHROMOSOMES, AS WELL AS SOME TRAITS THAT ARE SPECIFICALLY SEX-RELATED.
AND HE SAW THAT THEY FOLLOWED MENDEL'S RULES.
TOO BAD THAT MENDEL, THE MAN WHO JUMP-STARTED GENETIC SCIENCE, NEVER HIMSELF KNEW OF CHROMOSOMES AND GENES -- THE PHYSICAL AGENTS THAT ARE THE CAUSE BEHIND THE EFFECTS HE SO CAREFULLY OBSERVED!

The cell cycle-- MITOSIS.
I like it. It's stable and predictable-- like the tides.
Like US.
YES, MITOSIS IS THE BEST GUARANTEE THAT EVERY NEW GENERATION WILL REMAIN THE SAME.
VERY FEW VARIABLES CAN ENTER INTO THE PROCESS. THERE IS LITTLE OPPORTUNITY FOR CHANGE--
--FOR MUTATION...
...AND OUR SIMILARLY STABLE REPRODUCTIVE STRATEGY HAS FOR AGES WORKED VERY WELL FOR SQUINCHDOM.
BUT THAT SAME STABILITY CAN BE A LIABILITY WHEN UNSTABLE ENVIRONMENTAL FACTORS CALL FOR A RAPID, ADAPTIVE RESPONSE.
Such as the... GENETIC DIFFICULTIES we're now facing...
YES.
THE SOLUTION MIGHT JUST BE A REPRODUCTIVE PROCESS THAT GIVES SEXUAL ORGANISMS AN ADVANTAGE WE LACK...
...MEIOSIS! IT'S THE SPECIALIZED PROCESS THAT DERIVES FROM MITOSIS BUT, UNLIKE MITOSIS, HALVES THE DNA IN SEX CELLS TO 23 CHROMOSOMES. THIS ALLOWS TWO PARENT SEX CELLS...
HI! I'M MOM...
...AND I'M DAD!
...TO EACH CONTRIBUTE 23 CHROMOSOMES AND FORM A NEW, COMPLETELY UNIQUE OFFSPRING WITH A FULL COMPLEMENT OF 46 CHROMOSOMES!
WE'RE GOING TO CALL HIM A ZYGOTE.
IT'S A FAMILY NAME.

REMEMBER-- THE 44 NON-SEX-RELATED HUMAN CHROMOSOMES ARE DIPLOID. MEIOSIS DIVIDES EACH INTO TWO HAPLOID SETS...
IT'S LIKE I'M BEING TORN APART FROM MYSELF!
...WHILE SEPARATING THE TWO HAPLOID GONOSOMES-- THE SEX CHROMOSOMES-- AS WELL.
SO WHAT? I NEVER FELT WE HAD MUCH IN COMMON.
SO--
WHILE MITOSIS PRODUCES TWO CELLS WITH IDENTICAL DNA CONTENT...
...MEIOSIS PRODUCES FOUR CELLS, EACH WITH HALF THE FULL COMPLEMENT OF DNA.
MEIOSIS FURTHER ENHANCES GENETIC VARIABILITY BY RE-COMBINING ALLELES--
--IT EXCHANGES GENE VARIANTS ON HOMOLOGOUS CHROMOSOMES!
THIS I've got to see.
IN ADDITION TO ALL THAT, THE TWO INDIVIDUALS CONTRIBUTING THEIR SEX CELLS TO THE REPRODUCTIVE PROCESS MUST BE OF A DIFFERENT GENDER.
HI! I'M MOM...
...AND I'M DAD!
ONE MUST BE A FEMALE, WHOSE SEX CELLS ARE CALLED EGGS, AND THE OTHER A MALE, WHOSE SEX CELLS ARE CALLED SPERM.
Two genders--
Both contributing genetic material that's been halved and mixed around...
It's all so very-- ALIEN.

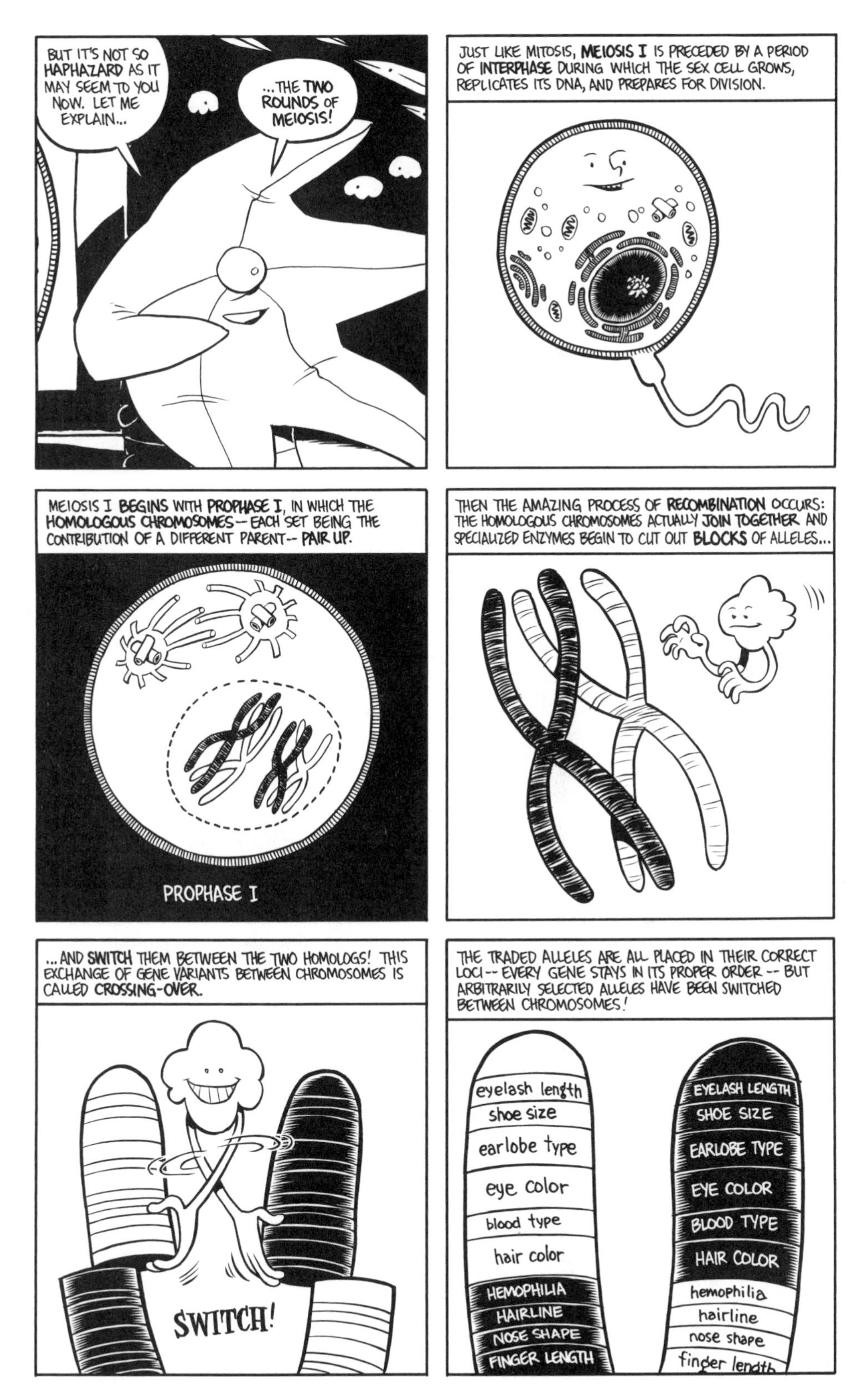
BUT IT'S NOT SO HAPHAZARD AS IT MAY SEEM TO YOU NOW. LET ME EXPLAIN...
...THE TWO ROUNDS OF MEIOSIS!
JUST LIKE MITOSIS, MEIOSIS I IS PRECEDED BY A PERIOD OF INTERPHASE DURING WHICH THE SEX CELL GROWS, REPLICATES ITS DNA, AND PREPARES FOR DIVISION.
MEIOSIS I BEGINS WITH PROPHASE I, IN WHICH THE HOMOLOGOUS CHROMOSOMES-- EACH SET BEING THE CONTRIBUTION OF A DIFFERENT PARENT-- PAIR UP.
PROPHASE I
THEN THE AMAZING PROCESS OF RECOMBINATION OCCURS: THE HOMOLOGOUS CHROMOSOMES ACTUALLY JOIN TOGETHER AND SPECIALIZED ENZYMES BEGIN TO CUT OUT BLOCKS OF ALLELES...
...AND SWITCH THEM BETWEEN THE TWO HOMOLOGS! THIS EXCHANGE OF GENE VARIANTS BETWEEN CHROMOSOMES IS CALLED CROSSING-OVER.
SWITCH!
THE TRADED ALLELES ARE ALL PLACED IN THEIR CORRECT LOCI-- EVERY GENE STAYS IN ITS PROPER ORDER -- BUT ARBITRARILY SELECTED ALLELES HAVE BEEN SWITCHED BETWEEN CHROMOSOMES!
eyelash length
shoe size
earlobe type
eye color
blood type
hair color
HEMOPHILIA
HAIRLINE
NOSE SHAPE
FINGER LENGTH
EYELASH LENGTH
SHOE SIZE
EARLOBE TYPE
EYE COLOR
BLOOD TYPE
HAIR COLOR
hemophilia
hairline
nose shape
finger length

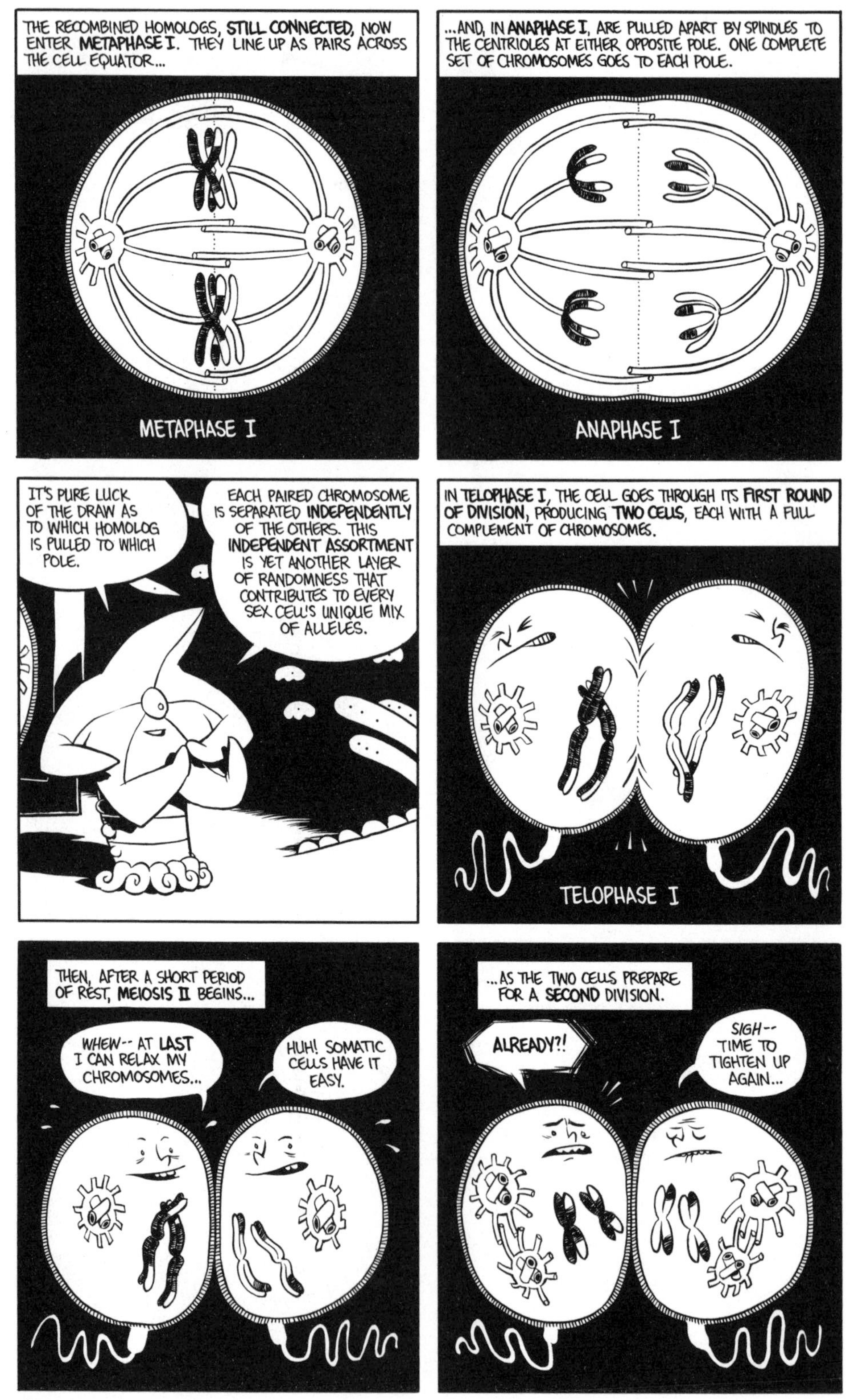
THE RECOMBINED HOMOLOGS, STILL CONNECTED, NOW ENTER METAPHASE I. THEY LINE UP AS PAIRS ACROSS THE CELL EQUATOR...
METAPHASE I
...AND, IN ANAPHASE I, ARE PULLED APART BY SPINDLES TO THE CENTRIOLES AT EITHER OPPOSITE POLE. ONE COMPLETE SET OF CHROMOSOMES GOES TO EACH POLE.
ANAPHASE I
IT'S PURE LUCK OF THE DRAW AS TO WHICH HOMOLOG IS PULLED TO WHICH POLE.
EACH PAIRED CHROMOSOME IS SEPARATED INDEPENDENTLY OF THE OTHERS. THIS INDEPENDENT ASSORTMENT IS YET ANOTHER LAYER OF RANDOMNESS THAT CONTRIBUTES TO EVERY SEX CELL'S UNIQUE MIX OF ALLELES.
IN TELOPHASE I, THE CELL GOES THROUGH ITS FIRST ROUND OF DIVISION, PRODUCING TWO CELLS, EACH WITH A FULL COMPLEMENT OF CHROMOSOMES.
TELOPHASE I
THEN, AFTER A SHORT PERIOD OF REST, MEIOSIS II BEGINS...
WHEW-- AT LAST I CAN RELAX MY CHROMOSOMES...
HUH! SOMATIC CELLS HAVE IT EASY.
...AS THE TWO CELLS PREPARE FOR A SECOND DIVISION.
ALREADY?!
SIGH-- TIME TO TIGHTEN UP AGAIN...

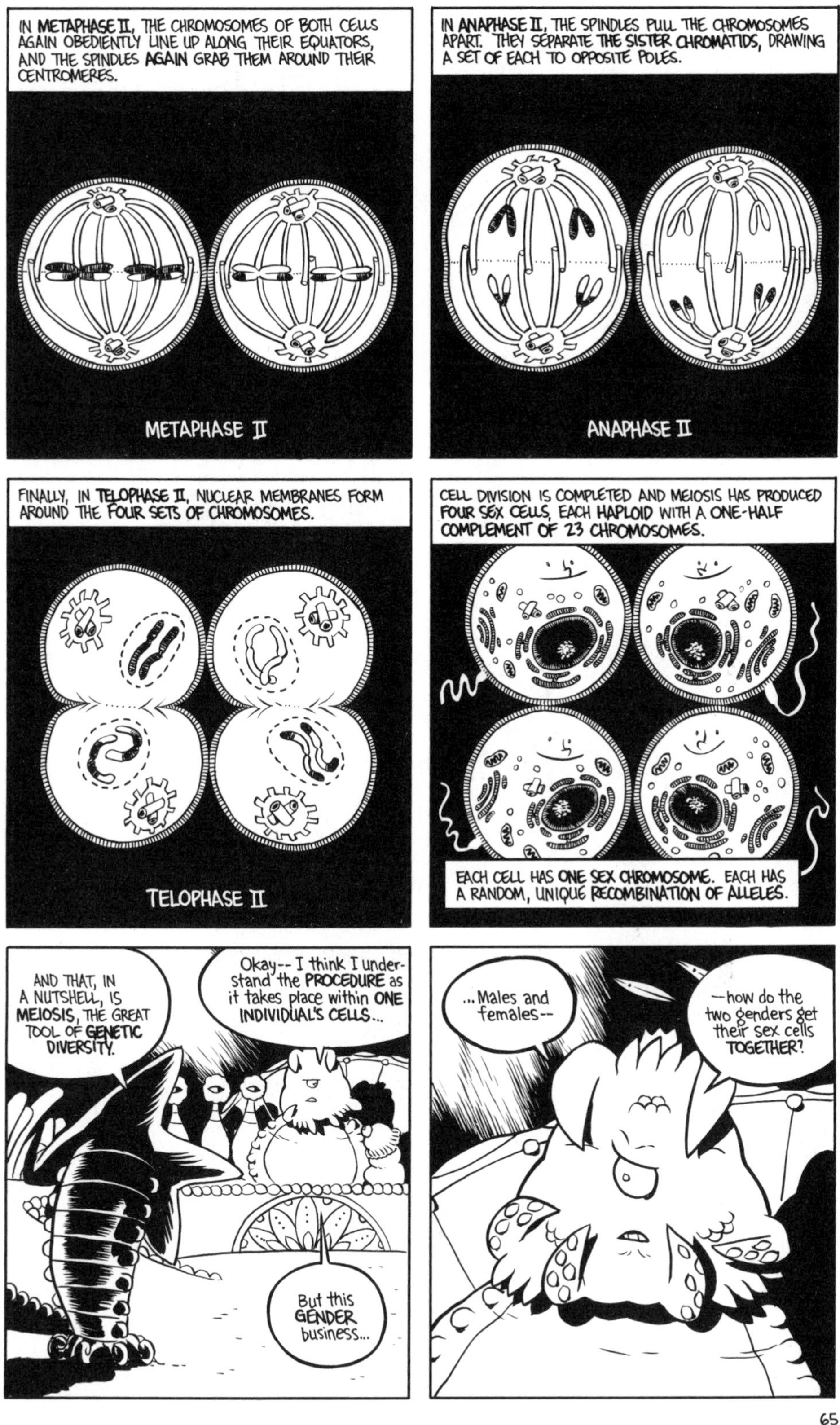

IN METAPHASE II, THE CHROMOSOMES OF BOTH CELLS AGAIN OBEDIENTLY LINE UP ALONG THEIR EQUATORS, AND THE SPINDLES AGAIN GRAB THEM AROUND THEIR CENTROMERES.
METAPHASE II
IN ANAPHASE II, THE SPINDLES PULL THE CHROMOSOMES APART. THEY SEPARATE THE SISTER CHROMATIDS, DRAWING A SET OF EACH TO OPPOSITE POLES.
ANAPHASE II
FINALLY, IN TELOPHASE II, NUCLEAR MEMBRANES FORM AROUND THE FOUR SETS OF CHROMOSOMES.
TELOPHASE II
CELL DIVISION IS COMPLETED AND MEIOSIS HAS PRODUCED FOUR SEX CELLS, EACH HAPLOID WITH A ONE-HALF COMPLEMENT OF 23 CHROMOSOMES.
EACH CELL HAS ONE SEX CHROMOSOME. EACH HAS A RANDOM, UNIQUE RECOMBINATION OF ALLELES.
AND THAT, IN A NUTSHELL, IS MEIOSIS, THE GREAT TOOL OF GENETIC DIVERSITY.
Okay-- I think I understand the PROCEDURE as it takes place within ONE INDIVIDUAL'S CELLS...
But this GENDER business...
...Males and females--
--how do the two genders get their sex cells TOGETHER?

WELL, THE PHYSICAL INTERACTION THAT INTRODUCES MALE AND FEMALE DNA TAKES ON MANY FORMS AND BEHAVIORAL VARIATIONS. SOME ARE QUITE-- AHEM-- SPECTACULAR.
BUT THAT'S A FIELD OF STUDY OUTSIDE THE SCOPE OF THIS REPORT--
SO I'LL KEEP MY EXPLANATION STRICTLY CELLULAR...
GAMETOGENESIS IS THE PROCESS THROUGH WHICH MALE AND FEMALE GAMETES --SEX CELLS-- ARE DEVELOPED AND MADE READY FOR FERTILIZATION--THE CREATION OF A NEW GENERATION!
HI! I'M MOM...
...AND I'M DAD!
MEIOSIS IS A PART OF GAMETOGENESIS. THE SEX CELLS THAT MEIOSIS PRODUCES IN MALES ARE CALLED SPERM, AND THOSE PRODUCED IN FEMALES ARE CALLED EGGS.
FOR GAMETOGENESIS, MALES PRODUCE SPECIAL CELLS, SPERMATOGONIA, IN THAT PART OF THEIR SEX ORGAN CALLED THE TESTES. THE SPERMATOGONIA EACH HAVE A FULL DIPLOID SET OF 46 CHROMOSOMES.
REMEMBER ME?
THROUGH MEIOSIS, THE SPERMATOGONIA ARE DIVIDED DOWN TO FOUR SPERM CELLS, EACH A HAPLOID WITH THE 23 CHROMOSOME HALF-SHARE.
NOW I'M... US!

AND BECAUSE THE COMBINATION OF SEX CHROMOSOMES THAT MAKES A HUMAN **MALE** ARE AN X AND A Y, MEIOSIS IN MALE GAMETES GUARANTEES THAT **ONE-HALF** THE SPERM PRODUCED WILL CARRY AN X AND **ONE-HALF** WILL CARRY A Y.
X
Y
X
Y
THE **FEMALE** PRODUCES HER DIPLOID **OOGONIA** CELL IN HER OVARIES. LIKE THE MALE EQUIVALENT, IT IS RENDERED DOWN, BY MEIOSIS, TO **FOUR HAPLOID EGGS**. BUT...
X
X
X
X
... **UNLIKE** IN SPERM PRODUCTION, ONLY **ONE** OF THE FOUR EGGS MATURES INTO A VIABLE REPRODUCTIVE CELL.
THAT'S OKAY, HONEY...
THIS IS BECAUSE OF THE HUGE AMOUNTS OF **CYTOPLASM** NEEDED TO NOURISH THE CELLULAR BEGINNINGS OF A NEW GENERATION. THE FEMALE STRATEGY IS TO **POOL** ITS CYTOPLASM INTO ONE WELL-STOCKED EGG, LEAVING THE OTHER THREE DORMANT.
...YOU GO ON WITHOUT US...
GO ASK YOUR MOTHER
AS ALREADY MENTIONED, VARIOUS **ENVIRONMENTAL, ANATOMICAL,** AND **SOCIAL ELEMENTS** DICTATE JUST **HOW AND WHEN** A MALE ORGANISM INTRODUCES HIS SPERM TO A FEMALE EGG...
...BUT ONCE THEY **ARE** INTRODUCED, WHEN THE CONDITIONS ARE **FAVORABLE**, ONE OF THE MANY DEPLOYED SPERM ENTERS THE SINGLE EGG AND **FERTILIZES** IT...
LATER FOR YOU, LOSERS!

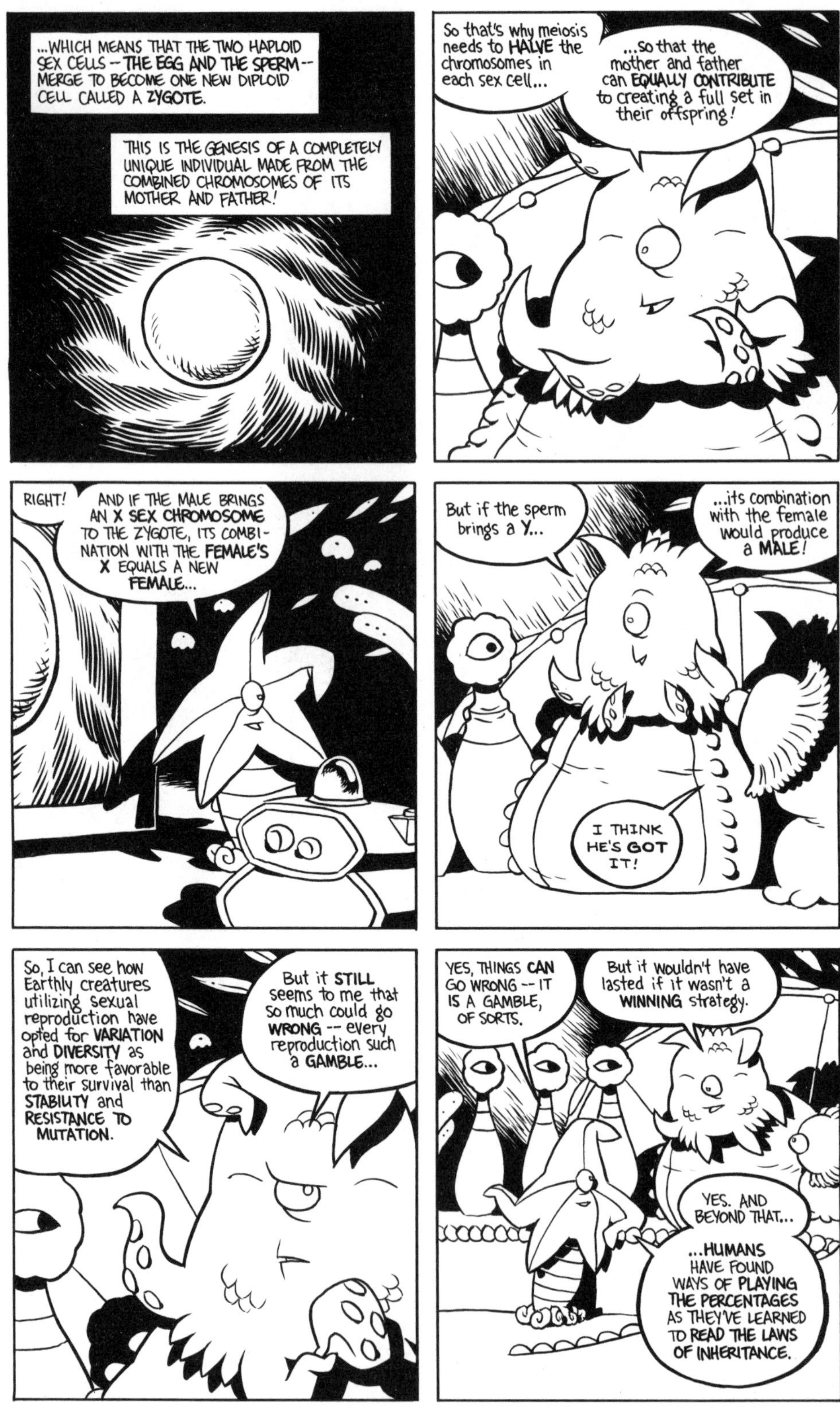
...WHICH MEANS THAT THE TWO HAPLOID SEX CELLS -- THE EGG AND THE SPERM -- MERGE TO BECOME ONE NEW DIPLOID CELL CALLED A ZYGOTE.
THIS IS THE GENESIS OF A COMPLETELY UNIQUE INDIVIDUAL MADE FROM THE COMBINED CHROMOSOMES OF ITS MOTHER AND FATHER!
So that's why meiosis needs to HALVE the chromosomes in each sex cell...
...so that the mother and father can EQUALLY CONTRIBUTE to creating a full set in their offspring!
RIGHT!
AND IF THE MALE BRINGS AN X SEX CHROMOSOME TO THE ZYGOTE, ITS COMBINATION WITH THE FEMALE'S X EQUALS A NEW FEMALE...
But if the sperm brings a Y...
...its combination with the female would produce a MALE!
I THINK HE'S GOT IT!
So, I can see how Earthly creatures utilizing sexual reproduction have opted for VARIATION and DIVERSITY as being more favorable to their survival than STABILITY and RESISTANCE TO MUTATION.
But it STILL seems to me that so much could go WRONG -- every reproduction such a GAMBLE...
YES, THINGS CAN GO WRONG -- IT IS A GAMBLE, OF SORTS.
But it wouldn't have lasted if it wasn't a WINNING strategy.
YES. AND BEYOND THAT...
...HUMANS HAVE FOUND WAYS OF PLAYING THE PERCENTAGES AS THEY'VE LEARNED TO READ THE LAWS OF INHERITANCE.

CHAPTER 3
How the System Works-- Everyone Gets an Inheritance
SO-- UP UNTIL NOW WE'VE BEEN DELVING INTO THE INCREDIBLY DIVERSE GENETIC POSSIBILITIES THAT COME WITH SEXUAL REPRODUCTION...
...THROUGH A LOOK AT THE MOLECULAR AND CELLULAR PROCESSES THAT MAKE THAT SYSTEM WORK.
I KNOW, I KNOW...
...NOT AS EXCITING AS THE PREVIOUS PANEL...
NOW WE'LL LOOK AT TRANSMISSION GENETICS--
--THE PATTERNS AND LAWS BY WHICH PHYSICAL CHARACTERISTICS ARE PASSED FROM ONE GENERATION TO THE NEXT...
LAWS?
I SWEAR WE'RE OVER 21 AND LEGALLY MARRIED!
...OTHERWISE KNOWN AS HEREDITY.
OH. RIGHT.
NEVER MIND.
AS WE'VE SEEN, THOSE WONDERFUL TRAIT VARIATIONS FOUND IN SEXUALLY REPRODUCING ORGANISMS ARE POSSIBLE BECAUSE A GENE CAN COME IN DIFFERENT VARIATIONS...
a
a
B
B
C
C
A
A
b
b
c
c
...CALLED ALLELES.

SWITCH!
WE'LL SEE HERE JUST HOW WHAT WE'VE LEARNED ABOUT THE CROSSING-OVER OF ALLELES IN THE MEIOSIS PROCESS IS OF PRACTICAL EFFECT ON SUCCESSIVE GENERATIONS.
THE PATTERNS IN WHICH TRAITS ARE INHERITED ARE PREDICTABLE. THE TRANSMISSION OF TRAITS THROUGH GENERATIONS CAN BE CALCULATED.
RS
RS
Rs
Rs
rS
rS
rs
rs
RRSS
RRsS
RRSs
rRSS
RRss
RrSS
rRsS
rRSs
RrsS
RrSs
rRss
rrSS
Rrss
rrsS
rrSs
THE DISCOVERY AND UNDERSTANDING OF THE LAWS OF INHERITANCE -- AND THEIR VARIATIONS -- HAVE GIVEN HUMANS A FIRM GRIP ON THEIR GENETIC HERITAGE.
EVEN AS AN OVERVIEW, IT'S HARD TO DISCUSS THIS PRACTICAL, APPLICABLE UNDERSTANDING OF THE RESULTS OF GENETICS WITHOUT INTRODUCING THE INSIGHTFUL HUMAN WHO FIRST RECOGNIZED THOSE LAWS...
...GREGOR MENDEL.
THE CONCEPT OF HERITABLE TRAITS PASSED FROM PARENT TO CHILD HAD, OF COURSE, BEEN OBSERVED AND UNDERSTOOD AT LEAST SINCE HUMANS FIRST BEGAN LIVING IN FAMILY GROUPS.
HE LOOKS JUST LIKE YOU!
...AND, NO DOUBT, THE ISSUE OF PATERNITY HAS ALWAYS BEEN OF GREAT IMPORTANCE FOR ANY SEXUAL ORGANISM RECOGNIZING THAT LIKE BEGETS LIKE...
HE LOOKS NOTHING LIKE ME!
...BUT, UNTIL MENDEL, ALL THEORIES OF HEREDITY WERE ESSENTIALLY JUST SHOTS IN THE DARK, UNSUPPORTED BY SOLID EVIDENCE. MENDEL WAS THE FIRST TO MAKE AN ACCURATE, SCIENTIFIC ANALYSIS OF HEREDITY BY MEANS OF PAINSTAKING, CONTROLLED OBSERVATION.
SO WHEN WILL THE PATERNITY TEST BE INVENTED?

BEFORE MENDEL

IT ALL STARTED WITH "**LIKE BEGETS LIKE.**" SOMEWHERE IN THEIR EARLY HISTORY, HUMANS RECOGNIZED THAT ORGANISMS COULD BE COUNTED ON TO PRODUCE MORE OF THE **SAME** ORGANISM -- AND **NOT** PRODUCE ANY OTHER KIND.

EARLY MAN WAS **PRACTICAL**. THE DESIRE TO DOMESTICATE ANIMALS AND PLANTS WENT HAND IN HAND WITH THE KNOWLEDGE THAT **SEX** AND **POLLINATION** PRODUCED **BIRTHS** AND **FERTILE FIELDS. SELECTIVE CROSSING AND BREEDING** BECAME TOOLS FOR STRENGTHENING VALUED TRAITS AND REMOVING UNBENEFICIAL ONES.

BUT THE "**FACTORS**" THAT ALLOWED FOR THE TRANSMISSION OF **LIKE TRAITS** FROM PARENT TO OFFSPRING COULD ONLY BE **GUESSED AT**, AND THE GUESSES WERE OFTEN CAUGHT UP IN SUPERSTITION AND CONTEMPORARY SOCIAL VALUES.

FOR CENTURIES, THE NOTION THAT **BLOOD** WAS THE TRANSMITTER OF HERITABLE TRAITS WAS WIDELY ACCEPTED AS FACT. THE CHILDREN OF BIRACIAL COUPLES OFTEN CARRIED FEATURES REMINISCENT OF BOTH PARENTS' ETHNICITY, AND THIS WAS SEEN AS EVIDENCE THAT THE TWO BLOODLINES HAD **MIXED**. THE ROYALTY OF EUROPE SUPPOSEDLY CARRIED A **SUPERIOR BLUISH BLOOD** IN THEIR VEINS AND BECAME REFERRED TO AS **BLUE BLOODS**. (ALL THE INBREEDING THAT ATTEMPTED TO KEEP THAT BLOOD BLUE DIDN'T DO THEM ANY FAVORS, THOUGH.)

PHILIP IV

THE ANCIENT **GREEKS** THOUGHT THAT PERHAPS THE REPRODUCTIVE MATERIAL WAS SECRETED FROM **ALL PARTS** OF THE ORGANISM EQUALLY, EACH PART RESPONSIBLE FOR PASSING ITS TRAITS TO THE OFFSPRING. THIS BECAME KNOWN AS **PANGENESIS** AND NO LESS A SCIENTIST THAN CHARLES DARWIN SUBSCRIBED TO A VERSION OF THE THEORY, PROPOSING THAT "**GEMMULES**" ARE COLLECTED FROM THROUGHOUT THE BODY DURING SEX, FOR TRANSMISSION TO THE NEXT GENERATION.

DARWIN

BUT NO THEORY WAS SUPPORTED BY HARD EVIDENCE UNTIL **GREGOR MENDEL** CONDUCTED HIS CAREFULLY CONTROLLED EXPERIMENTS IN 1865. HIS EXPERIMENTS ON GARDEN PEAS LED HIM TO ACCURATELY INTERPRET THE **NATURE** AND **ACTION** OF **GENES** -- THE NOW-ACCEPTED "**FACTORS**" OF INHERITANCE -- LONG BEFORE THEY WERE ACTUALLY SEEN OR GIVEN A NAME.

MENDEL

MENDEL WAS A 19th-CENTURY AUSTRIAN MONK WITH A PARTICULAR INTEREST IN THE PLANT VARIATIONS HE SAW IN HIS GARDENS. HE WAS ENCOURAGED TO STUDY THESE VARIATIONS BY COLLEAGUES IN BOTH RELIGION AND SCIENCE.
MENDEL MADE SOME VERY GOOD CHOICES WHILE DEVELOPING HIS SCIENTIFIC STUDY. HE CHOSE TO EXPERIMENT WITH THE COMMON PEA PLANT, WHICH PROVED TO BE AN EXCELLENT SUBJECT FOR UNCOVERING THE ACTIONS OF SIMPLE INHERITANCE.
THEN MENDEL MADE AN INTELLECTUAL BREAKTHROUGH THAT SEPARATED HIS STUDY FROM PREVIOUS ATTEMPTS TO UNDERSTAND HEREDITY: HE RECOGNIZED THAT PHYSICAL TRAITS MUST BE OBSERVED AND QUANTIFIED INDIVIDUALLY--NOT AS AN AGGREGATE OF TRAITS OR AS A GENERATION OF TRAITS.
HARD TO BELIEVE IT TOOK A HUMBLE MONK TO FIGURE THIS OUT.
HE MADE FORTUITOUS CHOICES IN THE SPECIFIC TRAITS HE DECIDED TO STUDY IN HIS PEA PLANTS, TOO. THEY WERE ALL THE RESULTS OF GENES EXPRESSING IN ONLY TWO ALLELIC VARIATIONS.
THE SHAPE OF THIS PEA IS ROUND...
...AND THIS ONE IS WRINKLED. THAT'S ALL--ROUND OR WRINKLED.
WE'LL SEE LATER THAT SOMETIMES GENES CAN COME IN MORE THAN TWO ALLELES. BUT, BY EITHER LUCK OR CAREFUL OBSERVATION, MENDEL LIMITED HIS EXPERIMENTS TO GENES THAT EXPRESSED IN ONLY TWO POSSIBLE PHENOTYPES...
RABBITS ARE NOT FOR ME--TOO CONFUSING!
...AND THIS KEPT HIS GROUNDBREAKING OBSERVATIONS RELATIVELY SIMPLE TO INTERPRET, WITH NONE OF THE PUZZLING COMPLEXITY THAT WOULD HAVE OCCURRED IF HE HAD CHOSEN TRAITS WITH MORE ALLELIC VARIATION.
ANY NEW SCIENCE STARTS WITH BABY STEPS...

THESE ARE THE SEVEN TRAITS THAT MENDEL CHOSE FOR HIS STUDIES ON HEREDITY IN PEA PLANTS, AND THE TWO PHENOTYPES IN WHICH EACH MIGHT BE EXPRESSED:
1 HEIGHT:
TALL OR SHORT
2 FLOWER LOCATION:
TIP OF STEM OR ALONG THE STEM
3 POD SHAPE:
INFLATED OR CONSTRICTED
4 POD COLOR:
GREEN OR YELLOW
5 SEED SHAPE:
ROUND OR WRINKLED
6 SEED COLOR:
YELLOW OR GREEN
7 SEED COAT COLOR:
GRAY OR WHITE
MENDEL INTERBRED HIS PEA PLANTS -- THOUSANDS OF THEM -- GENERATION AFTER GENERATION, AND BEGAN TO RECOGNIZE, CONFIRM, AND SUCCESSFULLY PREDICT CERTAIN PATTERNS OF TRAIT TRANSMISSION.
HE CAREFULLY CONTROLLED HIS EXPERIMENTS. PEA PLANTS ARE CAPABLE OF SELF-FERTILIZATION, AND TO ELIMINATE THAT POSSIBILITY HE TRIMMED HIS SUBJECT'S ANTHERS -- THE MALE SEX ORGAN. HE THEN CAREFULLY INTRODUCED POLLEN -- THE MALE GAMETE -- TO HIS PARENT PLANTS BY HAND.
BEYOND THE ADVANTAGE OF HAVING TO DEAL WITH ONLY TWO ALLELIC VARIATIONS PER TRAIT, THE GARDEN PEA WAS ALSO A GOOD CHOICE AS IT IS A TRUE-BREEDER.
WHEN A TRUE-BREEDER IS ALLOWED TO SELF-FERTILIZE, ITS OFFSPRING INVARIABLY RETAIN ALL THE TRAITS DISPLAYED BY THE PARENT -- NO SURPRISES.
POLLEN
"LIKE BEGETS LIKE" IN THE PUREST SENSE.
AND ONE FINAL LUCKY BREAK FOR MENDEL, SOMETHING HE COULD NOT HAVE KNOWN TO CONSIDER: ALL SEVEN OF HIS CHOSEN TRAITS ARE AUTOSOMAL -- MEANING THEY ARE LOCATED ON NON-SEX CHROMOSOMES.
SEX CHROMOSOME TRAITS, WHICH MAY EXPRESS DIFFERENTLY DEPENDING ON GENDER, COULD ONLY HAVE MUDDIED MENDEL'S NASCENT GENETIC STUDIES.
HEY, I'M A MONK, FOR HEAVEN'S SAKE!

MENDEL SET UP HIS PEA PLANT EXPERIMENTS SO HE COULD OBSERVE ALL SEVEN TRAITS INDIVIDUALLY. TO DO THIS, HE CROSSED TRUE-BREEDERS THAT DIFFERED BY ONLY ONE TRAIT -- FOR INSTANCE, HE WOULD CROSS A TALL PEA PLANT WITH A SHORT, WITH THE SIX OTHER TRAITS CONSISTENT BETWEEN PARENTS.
NO DIFFERENCE OTHER THAN THE HEIGHT...
HIS RESULTS OF THESE MONOHYBRID CROSSES WERE EXTREMELY CONSISTENT. IN THIS CASE, THE TALL CROSSED WITH THE SHORT ALWAYS PRODUCED TALL OFFSPRING PLANTS.
HMM...
THAT WAS AN IMPORTANT OBSERVATION: UP UNTIL THEN, THE ACCEPTED THEORY HELD THAT INHERITED TRAITS WERE A BLEND OF BOTH PARENTS' TRAITS. THERE WAS CLEARLY NO BLENDING OF HEIGHT TRAITS IN THE PEA PLANTS.
WHAT MENDEL DID SEE WAS THAT IN THE OFFSPRING -- CALLED AN F_1 GENERATION -- THE TALL PARENT PEA PLANT'S TALLNESS CONSISTENTLY EXPRESSED WHILE THE SHORT PEA PLANT'S SHORTNESS NEVER EXPRESSED.
...WHAT HAPPENED TO THE SHORT TRAIT?!
MENDEL DIDN'T STOP HIS EXPERIMENTS HERE. HE LET THE F_1 GENERATION SELF-FERTILIZE, AND OBSERVED SOMETHING UNEXPECTED.
F_1 GENERATION
YES! DIY!
DO IT YOURSELF!
WHILE 75% OF THE OFFSPRING OF THE F_1 GENERATION SHOWED THE TALL TRAIT...
...25% CAME UP SHORT!
THE SHORT TRAIT HAD APPARENTLY SKIPPED THE F_1 GENERATION, REAPPEARING IN F_2.
AND WHEN MENDEL ALLOWED THE F_2 GENERATION TO SELF-FERTILIZE...
HMM, HMM, HMM...
...HE SAW THAT THE SHORT PEA PLANTS OF THAT GENERATION WERE TRUE-BREEDERS, PRODUCING ONLY MORE SHORTS, WHILE THE OFFSPRING OF THE TALLS BROKE DOWN IN A 3:1 RATIO -- THREE TALL OFFSPRING FOR EVERY SHORT.

FROM HIS OBSERVATIONS, MENDEL CONCLUDED THAT THE **UNKNOWN FACTORS** PASSING TRAITS FROM PARENT TO OFFSPRING CAME IN SETS OF **TWO VARIANTS** FOR EACH TRAIT.
IT'S CONSISTENTLY ONE OR THE OTHER!
PARENT
F_1
F_2
FURTHERMORE, MENDEL SAW THAT **ONE** OF THE TWO VARIANTS OF ANY GIVEN TRAIT TENDED TO **DOMINATE** THE OTHER, **RECESSIVE**, VARIANT. THE **DOMINANT** WOULD **CONCEAL** THE PRESENCE OF THE **RECESSIVE**, ALTHOUGH THE RECESSIVE VARIANT COULD **REAPPEAR** IN SUCCESSIVE GENERATIONS.
1 THE FACTOR FOR **TALL** ALWAYS DOMINATED THE FACTOR FOR **SHORT**.
2 FLOWERS **ALONG** THE STEMS ALWAYS DOMINATED FLOWERS AT THE **TIPS**.
3 **INFLATED** PODS ALWAYS DOMINATED **CONSTRICTED** PODS.
4 **GREEN** PODS ALWAYS DOMINATED **YELLOW** PODS.
5 **ROUND** SEEDS ALWAYS DOMINATED **WRINKLED** SEEDS.
6 **YELLOW** SEEDS ALWAYS DOMINATED **GREEN** SEEDS.
7 **GRAY** SEED COATS ALWAYS DOMINATED **WHITE** SEED COATS.
THE BIOLOGICAL MECHANICS UNDERPINNING THIS WOULD, OF COURSE, NOT BE UNDERSTOOD UNTIL MUCH LATER, BUT MENDEL'S **FACTOR VARIANTS** WERE, INDEED, ALLELES -- **GENES**!
HE ACCURATELY DEDUCED THAT THE TWO PARENTS **EACH CONTRIBUTE ONE ALLELE** TO THE NEXT GENERATION.
WISH I COULD COME UP WITH A GOOD NAME FOR THESE **FACTORS**...
IF BOTH ALLELES INHERITED BY THE OFFSPRING WERE THE DOMINANT **TALL** ALLELE (**T AND T**), THE OFFSPRING WOULD BE **TALL**.
IF ONE ALLELE INHERITED WAS FOR **TALL (T)** AND THE OTHER WAS FOR **SHORT (t)**, THE TALL T WOULD **DOMINATE** -- CONCEAL -- THE SHORT t AND WOULD EXPRESS AS **TALL**.
ONLY IF **BOTH** ALLELES INHERITED WERE FOR **SHORT (t AND t)** WOULD THE OFFSPRING EXPRESS AS **SHORT**.

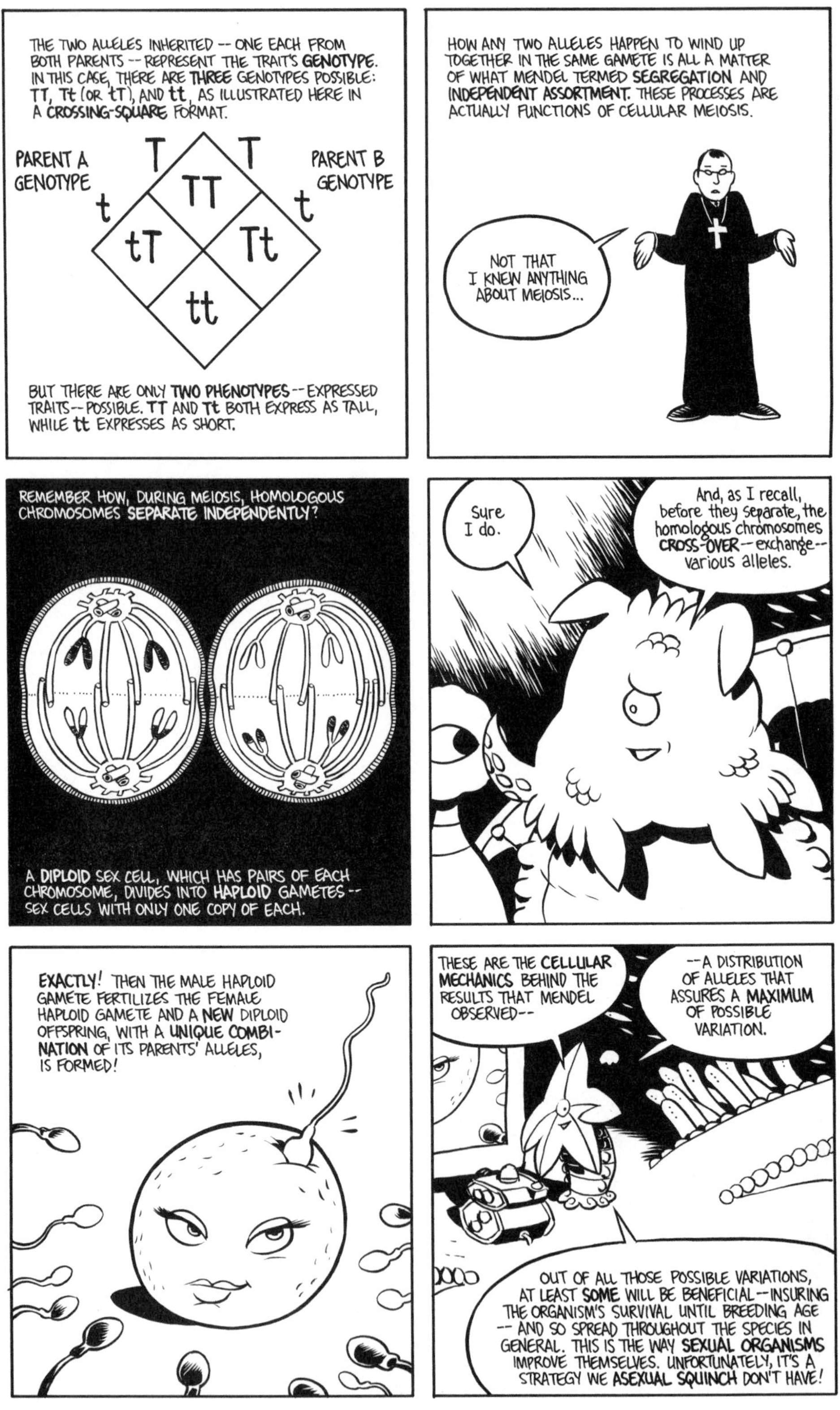
THE TWO ALLELES INHERITED -- ONE EACH FROM BOTH PARENTS -- REPRESENT THE TRAIT'S **GENOTYPE**. IN THIS CASE, THERE ARE **THREE** GENOTYPES POSSIBLE: **TT**, **Tt** (OR **tT**), AND **tt**, AS ILLUSTRATED HERE IN A **CROSSING-SQUARE** FORMAT.
PARENT A GENOTYPE
T
t
PARENT B GENOTYPE
T
t
TT
tT
Tt
tt
BUT THERE ARE ONLY **TWO PHENOTYPES** -- EXPRESSED TRAITS -- POSSIBLE. **TT** AND **Tt** BOTH EXPRESS AS TALL, WHILE **tt** EXPRESSES AS SHORT.
HOW ANY TWO ALLELES HAPPEN TO WIND UP TOGETHER IN THE SAME GAMETE IS ALL A MATTER OF WHAT MENDEL TERMED **SEGREGATION** AND **INDEPENDENT ASSORTMENT**. THESE PROCESSES ARE ACTUALLY FUNCTIONS OF CELLULAR MEIOSIS.
NOT THAT I KNEW ANYTHING ABOUT MEIOSIS...
REMEMBER HOW, DURING MEIOSIS, HOMOLOGOUS CHROMOSOMES **SEPARATE INDEPENDENTLY**?
A **DIPLOID** SEX CELL, WHICH HAS PAIRS OF EACH CHROMOSOME, DIVIDES INTO **HAPLOID** GAMETES -- SEX CELLS WITH ONLY ONE COPY OF EACH.
Sure I do.
And, as I recall, before they separate, the homologous chromosomes **CROSS-OVER** -- exchange -- various alleles.
EXACTLY! THEN THE MALE HAPLOID GAMETE FERTILIZES THE FEMALE HAPLOID GAMETE AND A **NEW** DIPLOID OFFSPRING, WITH A **UNIQUE COMBINATION** OF ITS PARENTS' ALLELES, IS FORMED!
THESE ARE THE **CELLULAR MECHANICS** BEHIND THE RESULTS THAT MENDEL OBSERVED--
--A DISTRIBUTION OF ALLELES THAT ASSURES A **MAXIMUM** OF POSSIBLE VARIATION.
OUT OF ALL THOSE POSSIBLE VARIATIONS, AT LEAST **SOME** WILL BE BENEFICIAL -- INSURING THE ORGANISM'S SURVIVAL UNTIL BREEDING AGE -- AND SO SPREAD THROUGHOUT THE SPECIES IN GENERAL. THIS IS THE WAY **SEXUAL ORGANISMS** IMPROVE THEMSELVES. UNFORTUNATELY, IT'S A STRATEGY WE **ASEXUAL SQUINCH** DON'T HAVE!

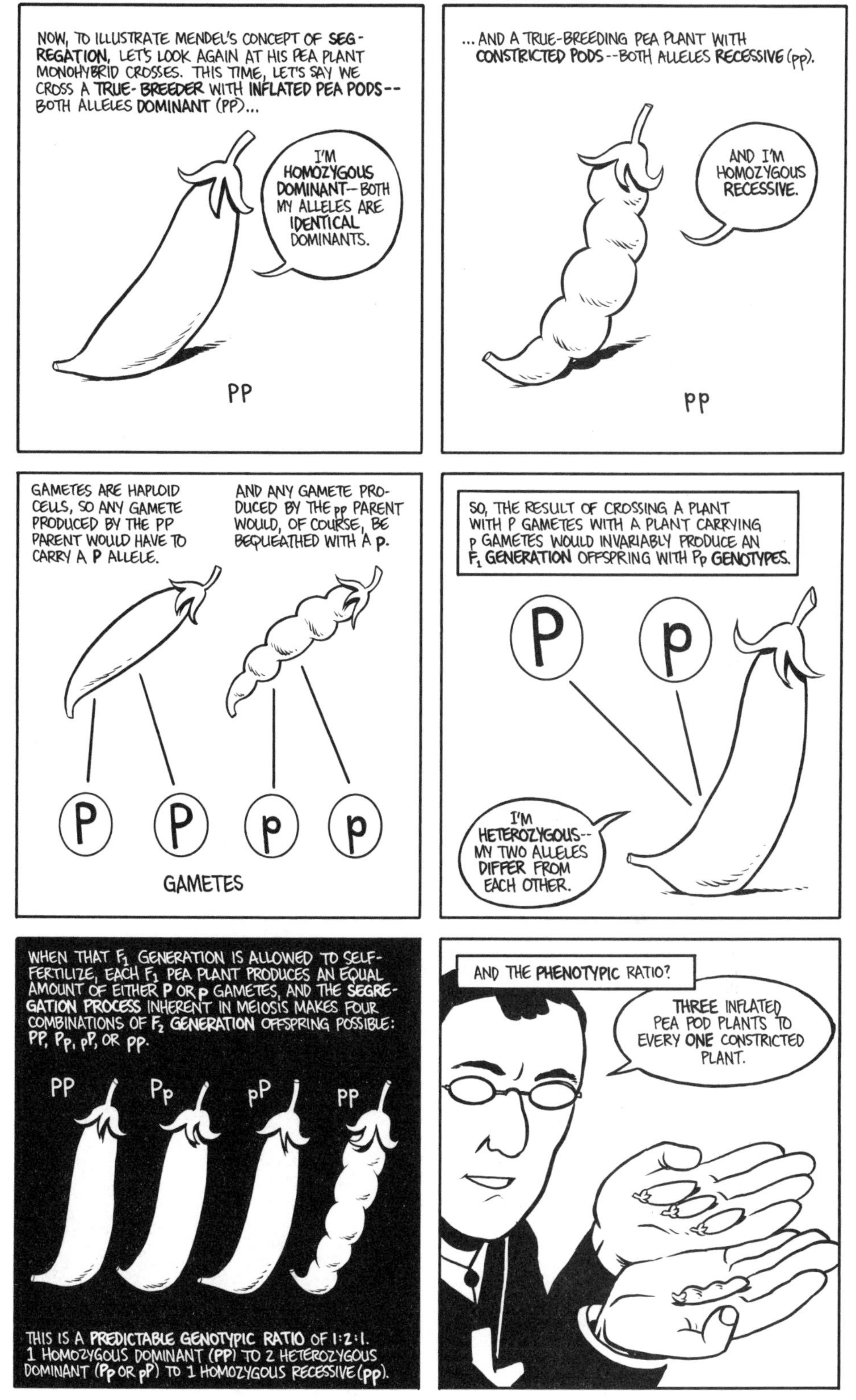
NOW, TO ILLUSTRATE MENDEL'S CONCEPT OF SEGREGATION, LET'S LOOK AGAIN AT HIS PEA PLANT MONOHYBRID CROSSES. THIS TIME, LET'S SAY WE CROSS A TRUE-BREEDER WITH INFLATED PEA PODS--BOTH ALLELES DOMINANT (PP)...
I'M HOMOZYGOUS DOMINANT--BOTH MY ALLELES ARE IDENTICAL DOMINANTS.
PP
...AND A TRUE-BREEDING PEA PLANT WITH CONSTRICTED PODS--BOTH ALLELES RECESSIVE (pp).
AND I'M HOMOZYGOUS RECESSIVE.
pp
GAMETES ARE HAPLOID CELLS, SO ANY GAMETE PRODUCED BY THE PP PARENT WOULD HAVE TO CARRY A P ALLELE.
AND ANY GAMETE PRODUCED BY THE pp PARENT WOULD, OF COURSE, BE BEQUEATHED WITH A p.
P P p p
GAMETES
SO, THE RESULT OF CROSSING A PLANT WITH P GAMETES WITH A PLANT CARRYING p GAMETES WOULD INVARIABLY PRODUCE AN F1 GENERATION OFFSPRING WITH Pp GENOTYPES.
P p
I'M HETEROZYGOUS--MY TWO ALLELES DIFFER FROM EACH OTHER.
WHEN THAT F1 GENERATION IS ALLOWED TO SELF-FERTILIZE, EACH F1 PEA PLANT PRODUCES AN EQUAL AMOUNT OF EITHER P OR p GAMETES, AND THE SEGREGATION PROCESS INHERENT IN MEIOSIS MAKES FOUR COMBINATIONS OF F2 GENERATION OFFSPRING POSSIBLE: PP, Pp, pP, OR pp.
PP Pp pP pp
THIS IS A PREDICTABLE GENOTYPIC RATIO OF 1:2:1. 1 HOMOZYGOUS DOMINANT (PP) TO 2 HETEROZYGOUS DOMINANT (Pp OR pP) TO 1 HOMOZYGOUS RECESSIVE (pp).
AND THE PHENOTYPIC RATIO?
THREE INFLATED PEA POD PLANTS TO EVERY ONE CONSTRICTED PLANT.

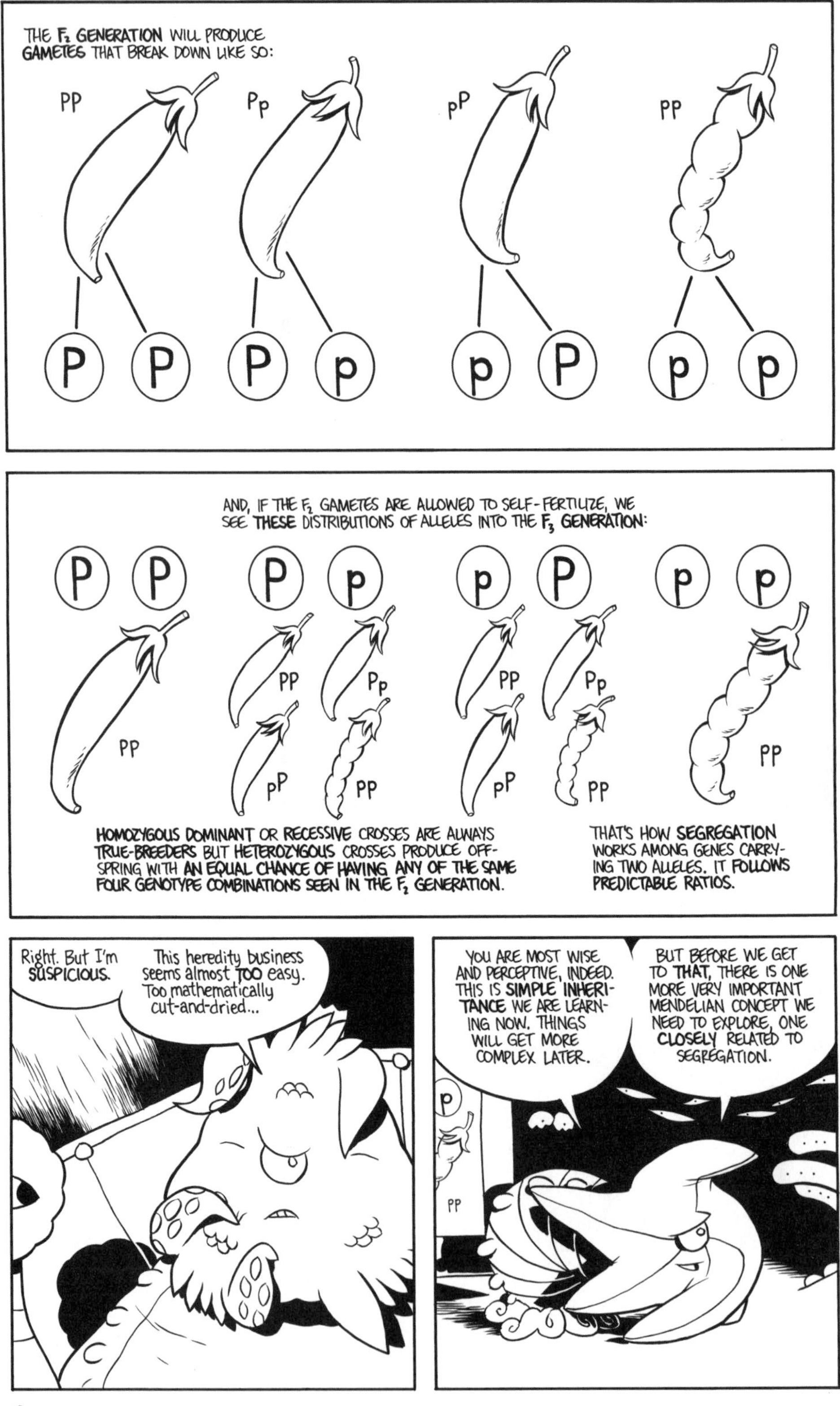
THE F_2 GENERATION WILL PRODUCE GAMETES THAT BREAK DOWN LIKE SO:
PP
Pp
pP
pp
P P P p p P p p
AND, IF THE F_2 GAMETES ARE ALLOWED TO SELF-FERTILIZE, WE SEE THESE DISTRIBUTIONS OF ALLELES INTO THE F_3 GENERATION:
P P P p p P p p
PP
PP Pp pP pp
PP Pp pP pp
pp
HOMOZYGOUS DOMINANT OR RECESSIVE CROSSES ARE ALWAYS TRUE-BREEDERS BUT HETEROZYGOUS CROSSES PRODUCE OFF-SPRING WITH AN EQUAL CHANCE OF HAVING ANY OF THE SAME FOUR GENOTYPE COMBINATIONS SEEN IN THE F_2 GENERATION.
THAT'S HOW SEGREGATION WORKS AMONG GENES CARRYING TWO ALLELES. IT FOLLOWS PREDICTABLE RATIOS.
Right. But I'm SUSPICIOUS.
This heredity business seems almost TOO easy. Too mathematically cut-and-dried...
YOU ARE MOST WISE AND PERCEPTIVE, INDEED. THIS IS SIMPLE INHERITANCE WE ARE LEARNING NOW. THINGS WILL GET MORE COMPLEX LATER.
BUT BEFORE WE GET TO THAT, THERE IS ONE MORE VERY IMPORTANT MENDELIAN CONCEPT WE NEED TO EXPLORE, ONE CLOSELY RELATED TO SEGREGATION.

INDEPENDENT ASSORTMENT ALLOWS OFFSPRING THE OPPORTUNITY TO INHERIT ANY POSSIBLE ALLELE FROM EITHER PARENT WITHOUT CONNECTION TO, OR INFLUENCE BY, ANY OTHER ALLELE.
EVERY ALLELE IS INHERITED INDEPENDENTLY.
TO INVESTIGATE THIS PRINCIPLE, WE NEED TO MOVE, LIKE MENDEL, BEYOND MONOHYBRIDS AND INTO THE REALM OF DIHYBRIDS--
--INDIVIDUALS THAT DIFFERENTIATE BY MORE THAN ONE TRAIT.
Please-- no more PEAS!
I've never even SEEN a REAL pea and already I'm tired of them!
OKAY-- FOR VARIETY AND SIMPLICITY'S SAKE, LET'S MAKE UP OUR OWN TEST SUBJECT.
LET'S MAKE IT AN ANIMAL-- A GARDEN INSECT-- THE VORACIOUS PEA BEETLE...
I SHOULD LOOK THIS HANDSOME IN REAL LIFE!
SO, TO BEGIN OUR DEMONSTRATION OF HOW TRAITS ARE SORTED INDEPENDENTLY OF ONE ANOTHER...
...WE'LL SAY OUR DIHYBRID SUBJECTS HAVE TWO ALLELIC VARIANCES--ONE IN CARAPACE SHAPE AND THE OTHER IN CARAPACE SPOT SIZE: A ROUND CARAPACE (R) IS DOMINANT OVER AN OBLONG CARAPACE (r) AND BIG SPOTS (S) ARE DOMINANT OVER LITTLE SPOTS (s).
CARAPACE...?
YOUR SHELL, SLOWPOKE-- YOUR SHELL!
BOTH OF OUR PEA BEETLES ARE HOMOZYGOUS FOR THEIR TRAITS. ONE HAS AN RRss GENOTYPE-- ROUND CARAPACE, LITTLE SPOTS. THE OTHER HAS AN rrSS GENOTYPE-- OBLONG CARAPACE, BIG SPOTS.
WE'RE SO DIFFERENT, YET I AM SO ATTRACTED TO YOU...
RRss
rrSS

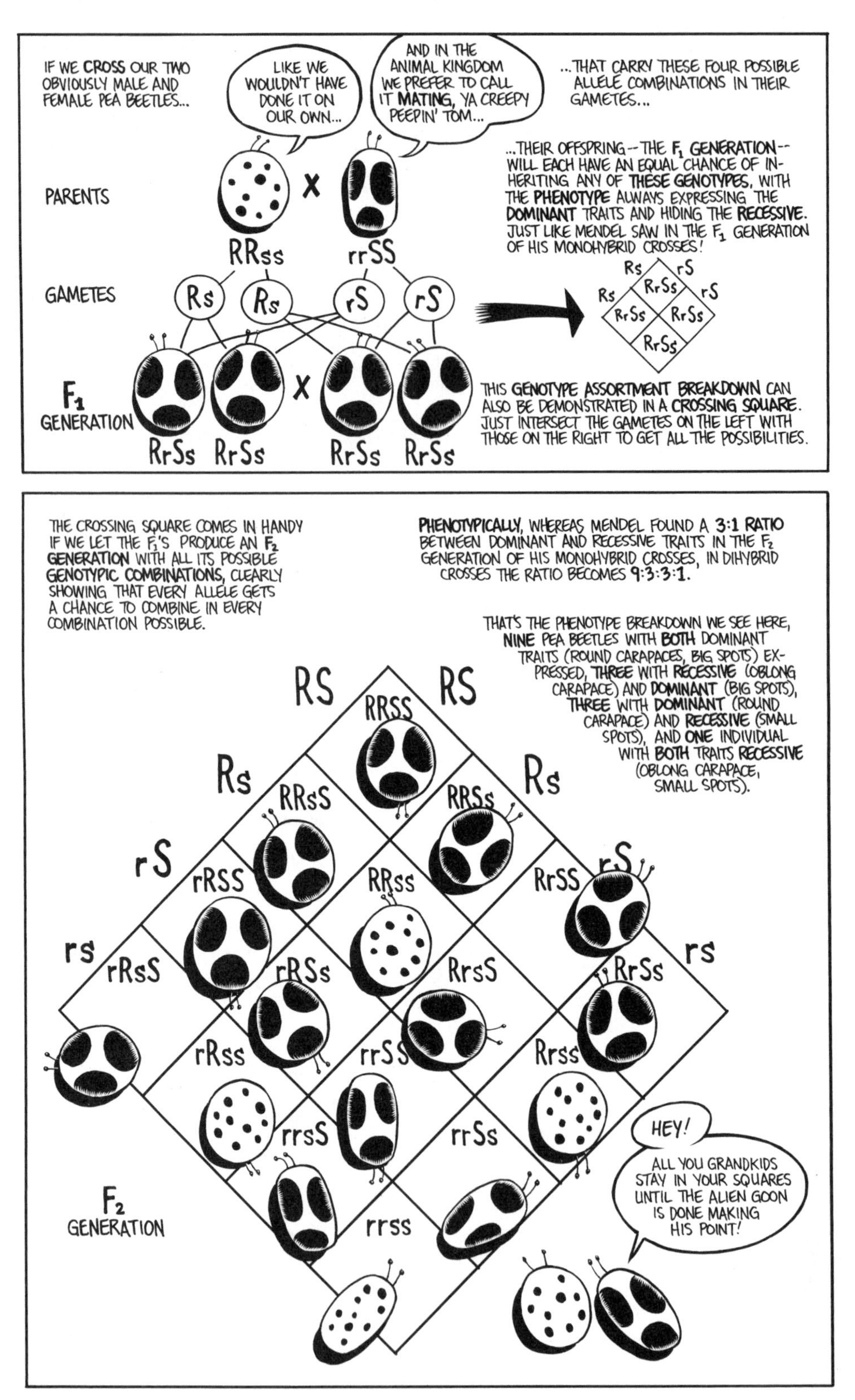

IF WE CROSS OUR TWO OBVIOUSLY MALE AND FEMALE PEA BEETLES...
LIKE WE WOULDN'T HAVE DONE IT ON OUR OWN...
AND IN THE ANIMAL KINGDOM WE PREFER TO CALL IT MATING, YA CREEPY PEEPIN' TOM...
...THAT CARRY THESE FOUR POSSIBLE ALLELE COMBINATIONS IN THEIR GAMETES...
...THEIR OFFSPRING--THE F1 GENERATION--WILL EACH HAVE AN EQUAL CHANCE OF INHERITING ANY OF THESE GENOTYPES, WITH THE PHENOTYPE ALWAYS EXPRESSING THE DOMINANT TRAITS AND HIDING THE RECESSIVE. JUST LIKE MENDEL SAW IN THE F1 GENERATION OF HIS MONOHYBRID CROSSES!
PARENTS
RRss
X
rrSS
GAMETES
Rs
Rs
rS
rS
Rs
rS
Rs
rS
RrSs
RrSs
RrSs
RrSs
F1 GENERATION
RrSs
RrSs
X
RrSs
RrSs
THIS GENOTYPE ASSORTMENT BREAKDOWN CAN ALSO BE DEMONSTRATED IN A CROSSING SQUARE. JUST INTERSECT THE GAMETES ON THE LEFT WITH THOSE ON THE RIGHT TO GET ALL THE POSSIBILITIES.
THE CROSSING SQUARE COMES IN HANDY IF WE LET THE F1'S PRODUCE AN F2 GENERATION WITH ALL ITS POSSIBLE GENOTYPIC COMBINATIONS, CLEARLY SHOWING THAT EVERY ALLELE GETS A CHANCE TO COMBINE IN EVERY COMBINATION POSSIBLE.
PHENOTYPICALLY, WHEREAS MENDEL FOUND A 3:1 RATIO BETWEEN DOMINANT AND RECESSIVE TRAITS IN THE F2 GENERATION OF HIS MONOHYBRID CROSSES, IN DIHYBRID CROSSES THE RATIO BECOMES 9:3:3:1.
THAT'S THE PHENOTYPE BREAKDOWN WE SEE HERE, NINE PEA BEETLES WITH BOTH DOMINANT TRAITS (ROUND CARAPACES, BIG SPOTS) EXPRESSED, THREE WITH RECESSIVE (OBLONG CARAPACE) AND DOMINANT (BIG SPOTS), THREE WITH DOMINANT (ROUND CARAPACE) AND RECESSIVE (SMALL SPOTS), AND ONE INDIVIDUAL WITH BOTH TRAITS RECESSIVE (OBLONG CARAPACE, SMALL SPOTS).
RS
RS
Rs
Rs
rS
rS
rs
rs
RRSS
RRsS
RRSs
rRSS
RRss
RrSS
rRsS
rRSs
RrsS
RrSs
rRss
rrSS
Rrss
rrsS
rrSs
rrss
F2 GENERATION
HEY!
ALL YOU GRANDKIDS STAY IN YOUR SQUARES UNTIL THE ALIEN GOON IS DONE MAKING HIS POINT!

HIS DIHYBRID EXPERIMENTS LED MENDEL TO ACCURATELY CONCLUDE THAT ALL TRAITS SORT OUT INDEPENDENTLY.
ALTHOUGH ONE TRAIT MAY DOMINATE THE EXPRESSION OF ANOTHER TRAIT, THE GENOTYPE IS INHERITED INDEPENDENTLY, WITH NO ONE TRAIT AFFECTING ANOTHER.
ALL TRAITS HAVE AN EQUAL CHANCE OF BEING SORTED OUT AND COMBINED WITH ANY OTHER TRAIT.
BOTANY
AND--OF COURSE--IT'S OUR OLD FRIEND MEIOSIS THAT SETS THE STAGE FOR THESE RECOMBINATIONS!
I DO ALL THE HEAVY LIFTING!
THOSE, THEN, ARE THE BASIC PRINCIPLES THAT GOVERN INHERITANCE. MENDEL DISCOVERED THEM, AND THEY HAVE STOOD THE TEST OF TIME PRETTY WELL. TO RECAP...
ALTERNATE VERSIONS OF GENES CALLED ALLELES ARE RESPONSIBLE FOR VARIATIONS IN ANY SPECIFIC TRAIT--
Every individual inherits TWO ALLELES for every trait.
One comes from each parent where they RANDOMLY SEGREGATED from their sister allele during MEIOSIS.
ALSO DURING MEIOSIS, EVERY ALLELE IS SORTED INDEPENDENTLY OF ALL OTHER ALLELES IN THE PROCESS OF CROSSING-OVER.
EVERY ALLELIC COMBINATION IS POSSIBLE.
AND, if the two alleles inherited by an individual are different, then ONE phenotype will always dominate the OTHER!

VERY GOOD, YOUR GALACTIC ASTUTENESS!
THOSE ARE THE PRINCIPLES THAT HAVE FORMED THE BEDROCK OF HUMAN UNDERSTANDING OF GENETICS. THEY MAKE PATTERNS OF INHERITANCE PREDICTABLE.
BUT...
...This is where you say BUT and let me know that this inheritance business isn't as CUT-AND-DRIED as I'd wish it to be...
RIGHT AGAIN, MY MOST SAGACIOUS LORDSHIP.
MENDEL'S OBSERVATIONS WERE SOLID -- BUT HIS CAREFUL CHOICE OF SUBJECT AND TRAITS INVOLVED SIMPLE INHERITANCE ONLY. THAT WAS JUST A START.
HUMANS HAVE SINCE LEARNED THAT INHERITANCE ISN'T ALWAYS SIMPLE. THERE ARE IMPORTANT EXCEPTIONS TO MENDEL'S OBSERVATIONS.
That figures. There's NOTHING simple about genetics...
WELL, IT'S NOT EASY, BUT IF YOU GET THE BASIC CONCEPTS, THE MORE COMPLEX STUFF ALL MAKES SENSE.
IN COMPLEX INHERITANCE CASES, IT'S IMPORTANT TO REALIZE THAT THE EXCEPTIONS TO MENDEL'S PRINCIPLES ALL INVOLVE HOW PHENOTYPES -- PHYSICAL TRAITS -- EXPRESS. SEGREGATION AND INDEPENDENT ASSORTMENT, THE LAWS GOVERNING THE INHERITANCE OF GENOTYPE, REMAIN CONSTANT.
I FIGURED THEY WOULD.
FOR INSTANCE, SIMPLE INHERITANCE DOES NOT ACCOMMODATE THE FACT THAT CERTAIN GENES DO NOT FOLLOW THE STRICT "DOMINANT MASKS RECESSIVE" FORM OF EXPRESSION.
THESE GENES EXPRESS IN DIFFERENT DEGREES OF DOMINANCE, RESULTING IN A PHENOTYPE SOMEWHERE BETWEEN THE DOMINANT AND RECESSIVE VARIATIONS...
AND IN MY DEFENSE I MUST SAY I NEVER CLAIMED THAT MY PRINCIPLES APPLIED TO ANYTHING OTHER THAN MY PEA PLANTS.

SNAPDRAGON FLOWERS ARE A GOOD EXAMPLE OF THIS INCOMPLETE DOMINANCE. FOLLOWING MENDEL'S EXPERIMENTS, IF WE CROSS A DOMINANT PURPLE-COLORED SNAPDRAGON (PP) WITH A RECESSIVE WHITE SNAPDRAGON (pp), WE WOULD EXPECT THE F_1 OFFSPRING (Pp) TO ALL EXPRESS THE DOMINANT PURPLE COLOR.
X
PP
pp
BUT, INSTEAD, WHAT WE GET IN SNAPDRAGON F_1 OFFSPRING IS A LIGHT PURPLE FLOWER THAT APPEARS TO BE THE RESULT OF A BLEND OF THE PURPLE AND WHITE COLOR ALLELES!
THIS INCOMPLETE DOMINANCE CONTINUES DOWN THE GENERATIONS, WITH THE F_2 POSSIBILITIES BREAKING DOWN IN THE 1:2:1 RATIO -- 1/4 PURPLE (PP), 1/2 LIGHT PURPLE (Pp OR pP), AND 1/4 WHITE (pp).
PP
Pp
pP
pp
HETEROZYGOUS (Pp, pP) OFFSPRING ALWAYS DISPLAY INCOMPLETE DOMINANCE, WITH LIGHT PURPLE FLOWERS THE RESULT.
BUT THE TWO ALLELES ARE NOT BLENDING THEIR TRAITS. GENOTYPICALLY, THEY CONTINUE TO OBSERVE THE LAWS OF SEGREGATION AND INDEPENDENT ASSORTMENT. THE DISTRIBUTION OF ALLELES DOWN THROUGH THE GENERATIONS REMAINS AS PREDICTABLE AS MENDEL'S FINDINGS.
OF COURSE I OBEY THE LAWS! I JUST LIKE TO DRESS MYSELF A BIT DIFFERENTLY...
ONLY THE PHENOTYPE DIFFERS, CAUSED BY CELLULAR MECHANICS WORKING OUT THE EXPRESSION OF THE COLOR PURPLE IN A DIFFERENT, MORE COMPLEX MANNER.
IT'S NOT HARD TO UNDERSTAND, HOWEVER, HOW PRE-MENDELIAN GENETICS MIGHT HAVE SEEN EXAMPLES OF INCOMPLETE DOMINANCE AS EVIDENCE FOR THE NOTION THAT THE TRAITS OF MOTHER AND FATHER WERE BLENDED TOGETHER IN THE OFFSPRING.
I MEAN, HOW OBVIOUS CAN THE EVIDENCE FOR HEREDITARY BLENDING BE?

ANOTHER VARIATION IN GENE EXPRESSION INVOLVES CODOMINANCE...
...WHICH OCCURS WHEN ALLELES SHARE EQUALLY IN THE EXPRESSION OF PHENOTYPE. BOTH ALLELES EXPRESS FULLY, TOO-- NO COMPROMISES!
MY TABBY COAT IS MADE UP OF BLACK HAIR ALLELES AND TAN HAIR ALLELES BOTH EXPRESSING COMPLETELY!
HUMAN BLOOD TYPES ARE A VERY WELL-KNOWN EXAMPLE OF CODOMINANCE.
BLOOD COMES IN DIFFERENT TYPES BECAUSE OF PROTEINS CALLED ANTIGENS THAT LIE ON THE SURFACE OF BLOOD CELLS, HELPING TO PROTECT HUMANS FROM DISEASE. THERE IS MORE THAN ONE KIND OF ANTIGEN...
...AND SEVERAL DIFFERENT ALLELES ARE RESPONSIBLE FOR PRODUCING THEM. THE MOST WIDESPREAD TYPES ARE CALLED A AND B-- ALLELES CODING FOR THESE ARE BOTH DOMINANT.
WE'RE THE A TEAM! WE DOMINATE!
WE'RE THE B TEAM! WE DOMINATE!
A
A
B
B
BUT SOME HUMANS CARRY TYPE AB BLOOD, A SITUATION WHERE BOTH ANTIGENS ARE PRESENT BECAUSE BOTH DOMINANT ALLELES HAVE SIMULTANEOUSLY AND FULLY EXPRESSED!
THE AB GENOTYPE EQUALS AN AB PHENOTYPE!
TOGETHER, WE CAN DOMINATE EVEN BETTER!
A
B
CODOMINANCE IS ALSO SEEN IN DOMESTIC CATTLE THAT SHOW THE REDDISH BROWN COLORATION CALLED ROAN.
ROAN OCCURS WHEN THE DOMINANT ALLELE FOR BROWN HAIR EXPRESSES EQUALLY WITH A DOMINANT ALLELE FOR WHITE.
THE RESULT OF THE TWO FULLY EXPRESSED ALLELES IS THE MIX OF HAIR COLORS THAT APPEARS REDDISH BROWN--ROAN.
BUT-- ONCE AGAIN-- THE CODOMINANCE VARIATION IS ALL IN THE PHYSICAL EXPRESSION OF THE TRAITS, THE PHENOTYPE. THE GENOTYPE STILL FOLLOWS THE RULES WE LEARNED FROM MENDEL.

NOW, SOMETIMES DOMINANT ALLELES DO NOT EXPRESS THEIR FULL INFLUENCE CONSISTENTLY.
THIS LACK OF TOTAL EXPRESSION IN THE PHENOTYPE IS CALLED INCOMPLETE PENETRANCE.
WHEN 100% OF INDIVIDUALS CARRYING A PARTICULAR DOMINANT ALLELE EXPRESS THE PHENOTYPE, THEN THE ALLELE IS SAID TO HAVE COMPLETE PENETRANCE.
MOST DOMINANT ALLELES ARE COMPLETELY PENETRANT.
INCOMPLETE PENETRANCE, HOWEVER, IS NOT UNCOMMON. AND A GOOD THING, TOO, IN THE CASE OF THIS EXAMPLE:
BREAST CANCER IS AN ALL-TOO-WIDESPREAD HUMAN SCOURGE THAT CAN BEGIN AS A MUTATION -- A SPONTANEOUS GENETIC CHANGE-- ARISING IN ONE OF A NUMBER OF DIFFERENT GENES AND THEN BE TRANSMITTED TO FUTURE GENERATIONS AS AN AUTOSOMAL DOMINANT TRAIT.
I'M ONE MUTATED CELL AND I'M JUST THE START!
THOSE MUTATED GENES-- DOMINANT ALLELES -- TEND TO DISPLAY INCOMPLETE PENETRANCE, HOWEVER.
ONLY A PERCENTAGE OF HUMANS CARRYING THIS POTENTIAL FOR BREAST CANCER IN THEIR GENOTYPE WILL ACTUALLY DEVELOP THE DISEASE ITSELF...
...ESTIMATES VARY BETWEEN 50 AND 85%.
ANOTHER, SLIGHTLY WEIRD EXAMPLE OF INCOMPLETE PENETRANCE: HUMANS ARE BILATERAL CREATURES NORMALLY CARRYING A COMPLEMENT OF FIVE DIGITS AT THE EXTREME ENDS OF EACH OF THEIR LIMBS. THEY CALL THEM FINGERS AND TOES.
IN CERTAIN HUMAN POPULATIONS, A TENDENCY TOWARD EXPRESSING MORE THAN FIVE DIGITS PER LIMB IS CARRIED IN THE GENOTYPE.
THIS IS, INTERESTINGLY ENOUGH, A DOMINANT TRAIT...
...WHICH, THANKFULLY FOR THOSE POPULATIONS, EXPRESSES WITH VERY INCOMPLETE PENETRANCE.

AS WE'VE ALREADY DISCUSSED, MENDEL WAS LUCKY--OR CLEVER--ENOUGH TO STUDY TRAITS THAT EXPRESSED IN ONLY TWO VARIATIONS.
I PREFER TO THINK OF MYSELF AS INSIGHTFUL.
BUT THE SITUATION CAN BECOME MUCH MORE COMPLICATED.
ALLELES CAN COME IN MORE THAN JUST TWO VARIATIONS...
a
a
A
A
B
B
b
b
C
C
c
c
...ALTHOUGH ANY INDIVIDUAL DIPLOID ORGANISM CAN, OF COURSE, HOLD ONLY TWO OF THOSE VARIATIONS IN ITS PARTICULAR GENOTYPE.
ALLELES--VARIANTS OF GENES--ARE, APPARENTLY, THE RESULTS OF MUTATIONS AND ARE GENERALLY BENEFICIAL TO THE ORGANISM.
THERE'S NO REASON WHY THERE WOULD BE A LIMIT ON THE NUMBER OF ALLELES THAT MIGHT POSSIBLY DEVELOP FROM ANY GIVEN GENE.
A VERY COMMON EXAMPLE OF THE INTERACTION OF MULTIPLE ALLELES IS A SMALL, FURRY DOMESTICATE CALLED THE RABBIT.
ITS WIDE VARIETY OF FUR COAT COLORS COMES COURTESY OF FOUR ALLELES THAT INTERACT AND FORM AN ARRANGEMENT CALLED A DOMINANCE HIERARCHY. HERE'S HOW IT WORKS...
THE COLOR BROWN (C+) IS CONSIDERED THE DEFAULT, "NORMAL" COLORATION--WHICH IS OFTEN REFERRED TO AS WILD TYPE, ON THE ASSUMPTION THAT THIS WAS THE COLORATION BEFORE DOMESTICATION AND MANIPULATIVE BREEDING.
MY BROWN ALLELE IS DOMINANT OVER THE OTHER THREE HAIR-COLOR ALLELES.
THE RECESSIVE ALBINO GENE (C) MUST BE HOMOZYGOUS--CARRY IDENTICAL ALLELES--TO EXPRESS. IT CODES FOR THE TOTAL ABSENCE OF COLOR, RESULTING IN WHITE FUR, AND PINK EYES AND SKIN.
BECAUSE THIS TRAIT IS ACTUALLY THE LACK OF A TRAIT (NO COLOR), ALBINISM IS THE RESULT OF WHAT'S KNOWN AS A NON-FUNCTIONING ALLELE.

THE CHINCHILLA ALLELE (C^{ch}) EXPRESSES AS A SOLID GRAY FUR COAT...
...AND IS PARTIALLY DOMINANT OVER HIMALAYAN!
THE FOURTH ALLELE-- HIMALAYAN (C^{h})-- SHOWS AS WHITE FUR WITH DARK EARS, NOSE, AND FEET.
CHINCHILLA/ HIMALAYAN CROSSES ARE HETEROZYGOUS AND EXPRESS AS GRAY FUR WITH DARK EARS, NOSE, AND FEET!

RABBIT FUR COLOR IS ACTUALLY DETERMINED BY FIVE DIFFERENT GENES. ALL THAT GENE INTERACTION, MULTIPLIED BY FOUR ALLELES, MAKES FOR A VERY COMPLICATED PICTURE!
BUT--AS ALWAYS-- MENDEL'S BASIC LAWS OF SEGREGATION AND INDEPENDENT ASSORTMENT STILL RULE THE GENOTYPE.

BY THE WAY-- THERE ARE SOME ALLELES THAT CARRY A DECIDEDLY UNBENEFICIAL TRAIT. LETHAL ALLELES ARE RECESSIVES THAT, WHEN HOMOZYGOUS, ALWAYS RESULT IN DEATH, OFTEN BEFORE BIRTH.

THERE ARE A FEW EXCEPTIONS: HUNTINGTON'S DISEASE, FOR INSTANCE, IS AN AUTOSOMAL DOMINANT LETHAL DISORDER THAT DOES NOT EXPRESS UNTIL HUMANS REACH ADULTHOOD.
EVEN SO, IT IS ALWAYS ULTIMATELY FATAL: 100% PENETRANCE.

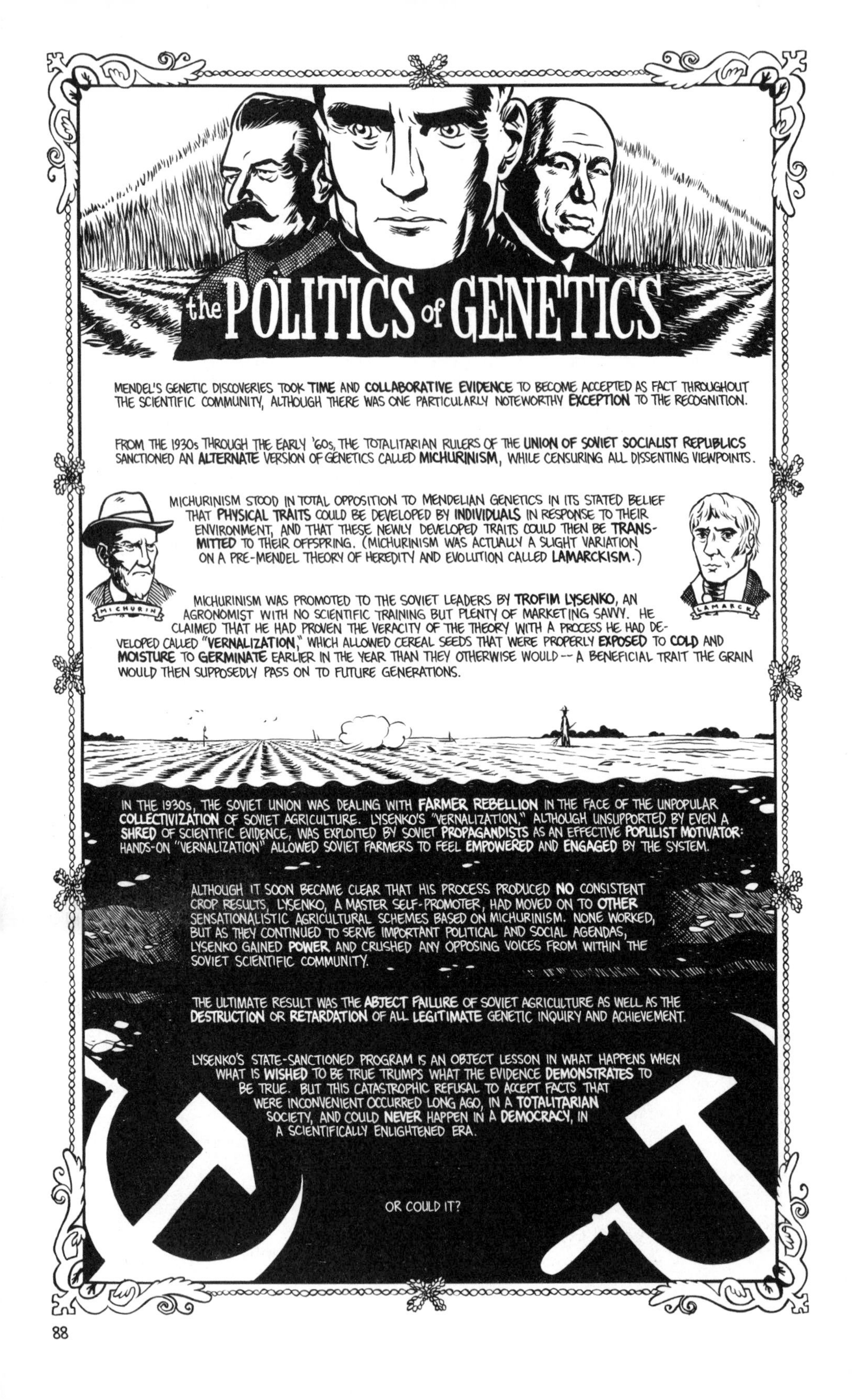
the POLITICS of GENETICS
MENDEL'S GENETIC DISCOVERIES TOOK TIME AND COLLABORATIVE EVIDENCE TO BECOME ACCEPTED AS FACT THROUGHOUT THE SCIENTIFIC COMMUNITY, ALTHOUGH THERE WAS ONE PARTICULARLY NOTEWORTHY EXCEPTION TO THE RECOGNITION.
FROM THE 1930s THROUGH THE EARLY '60s, THE TOTALITARIAN RULERS OF THE UNION OF SOVIET SOCIALIST REPUBLICS SANCTIONED AN ALTERNATE VERSION OF GENETICS CALLED MICHURINISM, WHILE CENSURING ALL DISSENTING VIEWPOINTS.
MICHURINISM STOOD IN TOTAL OPPOSITION TO MENDELIAN GENETICS IN ITS STATED BELIEF THAT PHYSICAL TRAITS COULD BE DEVELOPED BY INDIVIDUALS IN RESPONSE TO THEIR ENVIRONMENT, AND THAT THESE NEWLY DEVELOPED TRAITS COULD THEN BE TRANSMITTED TO THEIR OFFSPRING. (MICHURINISM WAS ACTUALLY A SLIGHT VARIATION ON A PRE-MENDEL THEORY OF HEREDITY AND EVOLUTION CALLED LAMARCKISM.)
MICHURIN
LAMARCK
MICHURINISM WAS PROMOTED TO THE SOVIET LEADERS BY TROFIM LYSENKO, AN AGRONOMIST WITH NO SCIENTIFIC TRAINING BUT PLENTY OF MARKETING SAVVY. HE CLAIMED THAT HE HAD PROVEN THE VERACITY OF THE THEORY WITH A PROCESS HE HAD DEVELOPED CALLED "VERNALIZATION," WHICH ALLOWED CEREAL SEEDS THAT WERE PROPERLY EXPOSED TO COLD AND MOISTURE TO GERMINATE EARLIER IN THE YEAR THAN THEY OTHERWISE WOULD -- A BENEFICIAL TRAIT THE GRAIN WOULD THEN SUPPOSEDLY PASS ON TO FUTURE GENERATIONS.
IN THE 1930s, THE SOVIET UNION WAS DEALING WITH FARMER REBELLION IN THE FACE OF THE UNPOPULAR COLLECTIVIZATION OF SOVIET AGRICULTURE. LYSENKO'S "VERNALIZATION," ALTHOUGH UNSUPPORTED BY EVEN A SHRED OF SCIENTIFIC EVIDENCE, WAS EXPLOITED BY SOVIET PROPAGANDISTS AS AN EFFECTIVE POPULIST MOTIVATOR: HANDS-ON "VERNALIZATION" ALLOWED SOVIET FARMERS TO FEEL EMPOWERED AND ENGAGED BY THE SYSTEM.
ALTHOUGH IT SOON BECAME CLEAR THAT HIS PROCESS PRODUCED NO CONSISTENT CROP RESULTS, LYSENKO, A MASTER SELF-PROMOTER, HAD MOVED ON TO OTHER SENSATIONALISTIC AGRICULTURAL SCHEMES BASED ON MICHURINISM. NONE WORKED, BUT AS THEY CONTINUED TO SERVE IMPORTANT POLITICAL AND SOCIAL AGENDAS, LYSENKO GAINED POWER AND CRUSHED ANY OPPOSING VOICES FROM WITHIN THE SOVIET SCIENTIFIC COMMUNITY.
THE ULTIMATE RESULT WAS THE ABJECT FAILURE OF SOVIET AGRICULTURE AS WELL AS THE DESTRUCTION OR RETARDATION OF ALL LEGITIMATE GENETIC INQUIRY AND ACHIEVEMENT.
LYSENKO'S STATE-SANCTIONED PROGRAM IS AN OBJECT LESSON IN WHAT HAPPENS WHEN WHAT IS WISHED TO BE TRUE TRUMPS WHAT THE EVIDENCE DEMONSTRATES TO BE TRUE. BUT THIS CATASTROPHIC REFUSAL TO ACCEPT FACTS THAT WERE INCONVENIENT OCCURRED LONG AGO, IN A TOTALITARIAN SOCIETY, AND COULD NEVER HAPPEN IN A DEMOCRACY, IN A SCIENTIFICALLY ENLIGHTENED ERA.
OR COULD IT?

THE SUBJECT OF LETHAL ALLELES CAN SERVE TO LEAD US INTO ANOTHER GENETIC PHENOMENON CALLED EPISTASIS...
ALWAYS GLAD TO BE OF ASSISTANCE...
...WHICH OCCURS WHEN CERTAIN GENES MASK THE EXPRESSION OF OTHER GENES.
OUR EXAMPLE ORGANISM WILL BE YET ANOTHER DOMESTICATED MAMMAL -- THE HORSE. HOW ITS HAIR COLOR EXPRESSES IS A GOOD EXAMPLE OF EPISTASIS AS WELL AS OF COMPLEX GENE INTERACTION.
ACTUALLY, THE GENETIC DETERMINATION OF HORSE HAIR COLOR IS SO COMPLICATED THAT OUR EXAMPLE HERE HAS BEEN GREATLY SIMPLIFIED FROM ALL THE INTER-ACTION THAT REALLY HAPPENS.
HORSES CARRY A DOMINANT ALLELE FOR WHITE (W) HAIR COLOR.
AS WITH RABBITS, W IS ACTUALLY A NONFUNCTIONING ALLELE FOR THE ABSENCE OF COLOR. BUT IN ITS HOMOZYGOUS (WW) FORM IT IS ALSO LETHAL.
WW
THERE ARE NO HORSES WITH THE WW GENOTYPE.
HORSES WITH THE HETEROZYGOUS Ww GENOTYPE HAVE A WHITE COAT.
ONLY THE RECESSIVE ww GENOTYPE ALLOWS OTHER GENES TO COME INTO PLAY AND VARIOUS COLORS TO BE EXPRESSED.
Ww
ww
THERE ARE ACTUALLY AT LEAST SEVEN GENES THAT INTERACT, DETERMINING THE MANY NUANCES OF COAT COLOR SEEN IN HORSES CARRYING THE ww GENOTYPE...
...MAKING HORSE BREEDING A KIND OF FINE ART!
I'M A HORSE OF A DIFFERENT COLOR!
BUT-- A DOMINANT W IN THE GENOTYPE HIDES THE ACTIONS OF ALL THE OTHER GENES RESPONSIBLE FOR HAIR COLORATION.
THIS IS DOMINANT EPISTASIS.
I'VE GOT LOTS OF COLOR IN ME...
...BUT I CAN'T GET IT OUT!

THERE ARE QUITE A FEW MORE VARIATIONS ON MENDEL'S OBSERVATIONS THAT LAYER EVEN MORE COMPLEXITY INTO GENE EXPRESSION.
GENES THAT ARE LOCATED CLOSE TO ONE ANOTHER ON THE CHROMOSOME ARE ALMOST ALWAYS LINKED-- INHERITED TOGETHER WITHOUT BENEFIT OF INDEPENDENT ASSORTMENT!
ONE GENE CAN HAVE A HAND IN MANY EXPRESSED TRAITS! THIS IS CALLED PLEIOTROPY, AND THOSE GENES ARE SAID TO BE PLEIOTROPIC.
AGING MAY BE A RESULT OF ANTAGONISTIC PLEIOTROPY -- GENES CREATING EFFECTS THAT BENEFIT US WHEN WE'RE YOUNG MAY CREATE OTHER EFFECTS THAT AGE US LATER IN LIFE!
THERE ARE TRAITS THAT GROW STRONGER AS THEY EXPRESS THROUGH SUCCESSIVE GENERATIONS! HUMANS ARE STILL TRYING TO FIGURE OUT THE MECHANICS OF THIS GENETIC ANTICIPATION!
THEY KEEP GETTING TALLER!
PARENT
F1
F2
AND WHILE MENDEL'S PEA PLANT TRAITS ALL CAME FROM GENES FOUND ON AUTOSOMAL CHROMOSOMES, THERE ARE PLENTY OF OTHER GENES ON THE SEX CHROMOSOMES -- MOSTLY ON THE X CHROMOSOME, ACTUALLY-- THAT EXPRESS DIFFERENTLY DEPENDING ON THE GENDER OF THE CARRIER!
THOMAS HUNT MORGAN DISCOVERED THAT FRUIT FLY EYE COLOR DIFFERENCES ARE CONTROLLED BY GENES ON THE Y CHROMOSOME! CHECK IT OUT!
WHEW! I COULD GO ON...
...But that's ENOUGH for now--I get the POINT...
...Mendel's law of inheritance will always prove a strong foundation for understanding genotype--but phenotype expression can be a bit more... UNRULY.
A GOOD ENOUGH SUMMATION, YOUR MOST PATIENT SCHOLARLINESS.
NOW LET'S TAKE A SHORT BREAK BEFORE WE CONTINUE WITH A LOOK AT WHAT HUMANS ARE ACTUALLY DOING WITH THEIR GROWING KNOWLEDGE OF GENETICS!

CHAPTER 4
Applying All That Stuff-- For the Greater Good
NOW THAT WE'VE COVERED THE BASIC MECHANICS OF GENETICS...
... IT'S TIME WE LOOKED AT THE PRACTICAL -- AND SOMETIMES IMPRACTICAL -- APPLICATIONS TO WHICH HUMANS HAVE PUT THEIR HARD-WON KNOWLEDGE.
LONG BEFORE MENDEL, EVEN THE AVERAGE ANIMAL BREEDER UNDERSTOOD THE BASICS OF PRACTICAL HEREDITY. STOCK ANIMALS WITH A SPECIFIC TRAIT BRED WITH OTHERS CARRYING THE SAME TRAIT INCREASED THE LIKELIHOOD THAT OFFSPRING WOULD INHERIT THAT TRAIT -- WHETHER BENEFICIAL OR HARMFUL.
AND IN SOME CASES, IT BECAME UNDERSTOOD THAT THE INTER-BREEDING OF STOCK TOO CLOSELY RELATED BY FAMILY TENDED TO INCREASE THE APPEARANCES OF UNHEALTHY TRAITS.
IT WAS SURELY NOTICED THAT SIMILAR UNHAPPY INHERITANCES COULD ALSO BEFALL HUMAN POPULATIONS.
HEMOPHILIA AND POLYDACTYLY ARE THE PRICES WE HAVE PAID TO REMAIN BLUE BLOODS!
BUT IT TOOK MENDEL AND HIS LAWS OF HEREDITY TO BEGIN THE SCIENTIFIC, SYSTEMATIC UNDERSTANDING OF EXACTLY HOW AND WHEN GENETIC DISORDERS ARE TRANSMITTED...
...AND, WITH THAT, THE HOPE THAT SOMETHING COULD BE DONE TO HELP GUIDE INDIVIDUALS PAST GENETIC DANGERS...
WELL, ACTUALLY, I WAS ONLY THINKING IN TERMS OF PEA PLANTS.
...SOMETHING THAT BECAME KNOWN AS GENETIC COUNSELING...
...THE SYSTEMATIC EXAMINATION OF FAMILY HISTORY FOR HEREDITARY DISORDERS AND THE DETERMINATION OF THE LIKELIHOOD OF A SPECIFIC FAMILY MEMBER INHERITING A DISORDER, COUPLED WITH PRACTICAL ADVICE CONCERNING THE DETERMINATION.
A LITTLE TOO LATE TO HELP US.

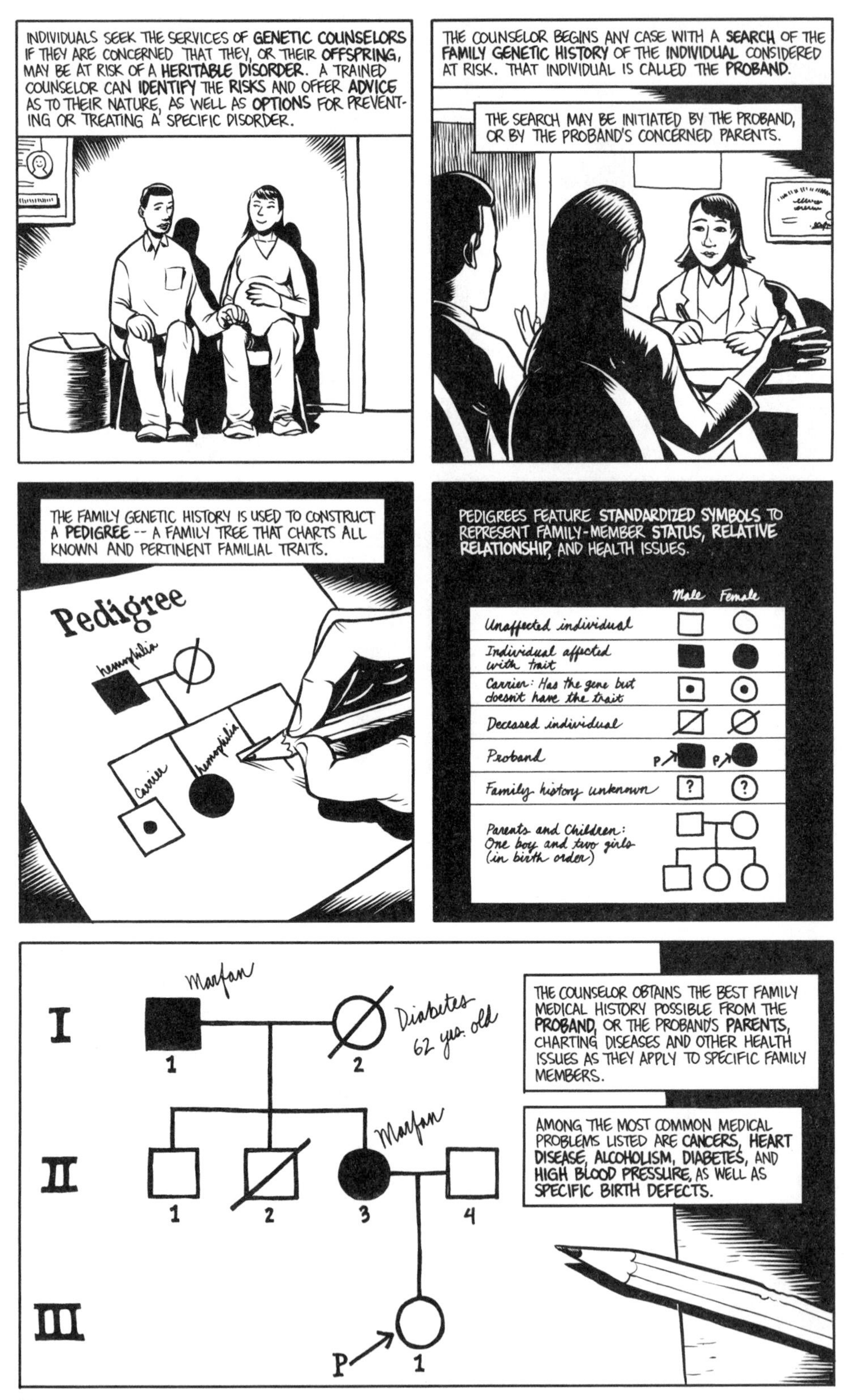

INDIVIDUALS SEEK THE SERVICES OF GENETIC COUNSELORS IF THEY ARE CONCERNED THAT THEY, OR THEIR OFFSPRING, MAY BE AT RISK OF A HERITABLE DISORDER. A TRAINED COUNSELOR CAN IDENTIFY THE RISKS AND OFFER ADVICE AS TO THEIR NATURE, AS WELL AS OPTIONS FOR PREVENTING OR TREATING A SPECIFIC DISORDER.
THE COUNSELOR BEGINS ANY CASE WITH A SEARCH OF THE FAMILY GENETIC HISTORY OF THE INDIVIDUAL CONSIDERED AT RISK. THAT INDIVIDUAL IS CALLED THE PROBAND.
THE SEARCH MAY BE INITIATED BY THE PROBAND, OR BY THE PROBAND'S CONCERNED PARENTS.
THE FAMILY GENETIC HISTORY IS USED TO CONSTRUCT A PEDIGREE -- A FAMILY TREE THAT CHARTS ALL KNOWN AND PERTINENT FAMILIAL TRAITS.
Pedigree
hemophilia
hemophilia
Carrier
PEDIGREES FEATURE STANDARDIZED SYMBOLS TO REPRESENT FAMILY-MEMBER STATUS, RELATIVE RELATIONSHIP, AND HEALTH ISSUES.
Male
Female
Unaffected individual
Individual affected with trait
Carrier: Has the gene but doesn't have the trait
Deceased individual
Proband
P
P
Family history unknown
?
?
Parents and Children: One boy and two girls (in birth order)
I
II
III
Marfan
Diabetes 62 yrs. old
1
2
Marfan
1
2
3
4
P
1
THE COUNSELOR OBTAINS THE BEST FAMILY MEDICAL HISTORY POSSIBLE FROM THE PROBAND, OR THE PROBAND'S PARENTS, CHARTING DISEASES AND OTHER HEALTH ISSUES AS THEY APPLY TO SPECIFIC FAMILY MEMBERS.
AMONG THE MOST COMMON MEDICAL PROBLEMS LISTED ARE CANCERS, HEART DISEASE, ALCOHOLISM, DIABETES, AND HIGH BLOOD PRESSURE, AS WELL AS SPECIFIC BIRTH DEFECTS.

AS SPECIFIC GENETIC TRAITS ARE IDENTIFIED WITHIN THE FAMILY...
...THE COUNSELOR'S EXPERTISE WITH MENDELIAN GENETICS HELPS DETERMINE THE CHANCES THAT THE PROBAND--
--OR ANY OTHER CONCERNED FAMILY MEMBER--
--MIGHT INHERIT THE TRAIT.
GENETIC COUNSELORS REFER TO ANY FAMILY MEMBER CARRYING ONE COPY OF THE GENE CODING FOR A SPECIFIC DISORDER AS HETEROZYGOUS...
YOU'RE GOING TO MAKE AN EXAMPLE OF OUR SON AGAIN, AREN'T YOU?
...WHILE ANY MEMBER POSSESSING TWO COPIES OF THE GENE CODING FOR A SPECIFIC DISORDER IS CALLED HOMOZYGOUS.
SOUND FAMILIAR? IT SHOULD-- IT'S THE SAME STANDARD MENDELIAN TERMINOLOGY THAT APPLIES TO PEA PLANTS.
HIS MOTHER AND I DIDN'T EXPRESS, BUT...
ANY FAMILY MEMBER EXPRESSING A SPECIFIC DISORDER IS TERMED AFFECTED.
ALTHOUGH OUR POLYDACTYLY IS A DOMINANT DISORDER, MY PARENTS ARE NOT AFFECTED BECAUSE IT IS ALSO INCOMPLETELY PENETRANT AND HAS NOT EXPRESSED IN THEM.
AND AS WE'VE LEARNED, DIFFERENT TRAITS, DISORDERS INCLUDED, ARE TRANSMITTED GENERATION TO GENERATION IN DIFFERENT PATTERNS-- MODES OF INHERITANCE-- DEPENDING ON DOMINANT AND RECESSIVE CHARACTERISTICS.
PARENT
F1
F2
HERITABLE TRAITS IN HUMANS FOLLOW THE SAME RULES OF TRANSMISSION AS DISCOVERED BY MENDEL-- WITH A FEW CURVES THROWN IN BY GENES LOCATED ON THE SEX CHROMOSOMES.
REMEMBER--
--FEMALES CARRY TWO X CHROMOSOMES AND NO Y, WHILE MALES HAVE A Y AND ONLY ONE X.

MODES of INHERITANCE
CHECKLIST
AUTOSOMAL DOMINANT
TRAITS ARE CODED BY GENES ON ANY NON-SEX CHROMOSOME
BOTH MALES AND FEMALES ARE AFFECTED EQUALLY (ALTHOUGH PENETRANCE MAY NOT ALWAYS BE COMPLETE)
TRAITS DO NOT SKIP GENERATIONS -- AFFECTED CHILDREN ARE BORN TO AFFECTED PARENTS
DISORDERS INCLUDE POLYDACTYLY, MARFAN SYNDROME, AND HUNTINGTON'S DISEASE
AUTOSOMAL RECESSIVE
TRAITS ARE CODED BY GENES ON ANY NON-SEX CHROMOSOME
TRAITS ARE EXPRESSED ONLY WHEN TWO IDENTICAL ALLELES ARE INHERITED
BOTH MALES AND FEMALES ARE AFFECTED EQUALLY
TRAITS CAN SKIP GENERATIONS -- AFFECTED CHILDREN MAY BE BORN TO UNAFFECTED PARENTS
CHILDREN BORN WITHIN A CLOSELY RELATED POPULATION HAVE INCREASED CHANCES OF INHERITING TRAITS
DISORDERS INCLUDE CYSTIC FIBROSIS AND TAY-SACHS DISEASE
X-LINKED RECESSIVE
TRAITS ARE ONLY CODED BY GENES ON FEMALE SEX CHROMOSOME
MALES ARE MOSTLY AFFECTED, AS THEY HAVE ONLY ONE X CHROMOSOME, WHICH CANNOT BE OFFSET
TRAITS SKIP GENERATIONS
AFFECTED SONS ARE BORN TO UNAFFECTED MOTHERS
TRAITS ARE NEVER PASSED FROM FATHER TO SON, BUT DAUGHTERS ARE ALWAYS CARRIERS
DISORDERS INCLUDE HEMOPHILIA AND DUCHENNE MUSCULAR DYSTROPHY
X-LINKED DOMINANT
TRAITS ARE ONLY CODED BY GENES ON FEMALE SEX CHROMOSOME
TRAITS DON'T SKIP GENERATIONS; BOTH MALES AND FEMALES ARE AFFECTED
AFFECTED MOTHERS HAVE BOTH AFFECTED SONS AND AFFECTED DAUGHTERS
AFFECTED FATHERS ALWAYS HAVE AFFECTED DAUGHTERS
DISORDERS INCLUDE INCONTINENTIA PIGMENTI AND FRAGILE X SYNDROME
Y-LINKED
TRAITS ARE ONLY CODED BY GENES ON MALE SEX CHROMOSOME
TRAITS ARE EXCLUSIVELY PASSED FROM FATHER TO SON; TRAITS DON'T SKIP GENERATIONS
MOST TRAITS ARE RELATED TO MALE SEXUAL CHARACTERISTICS
DISORDERS SUCH AS SCHIZOPHRENIA AND PARKINSON'S DISEASE MAY BE LINKED TO THE SRY GENE
AND HAIRY EARS ARE Y-LINKED!

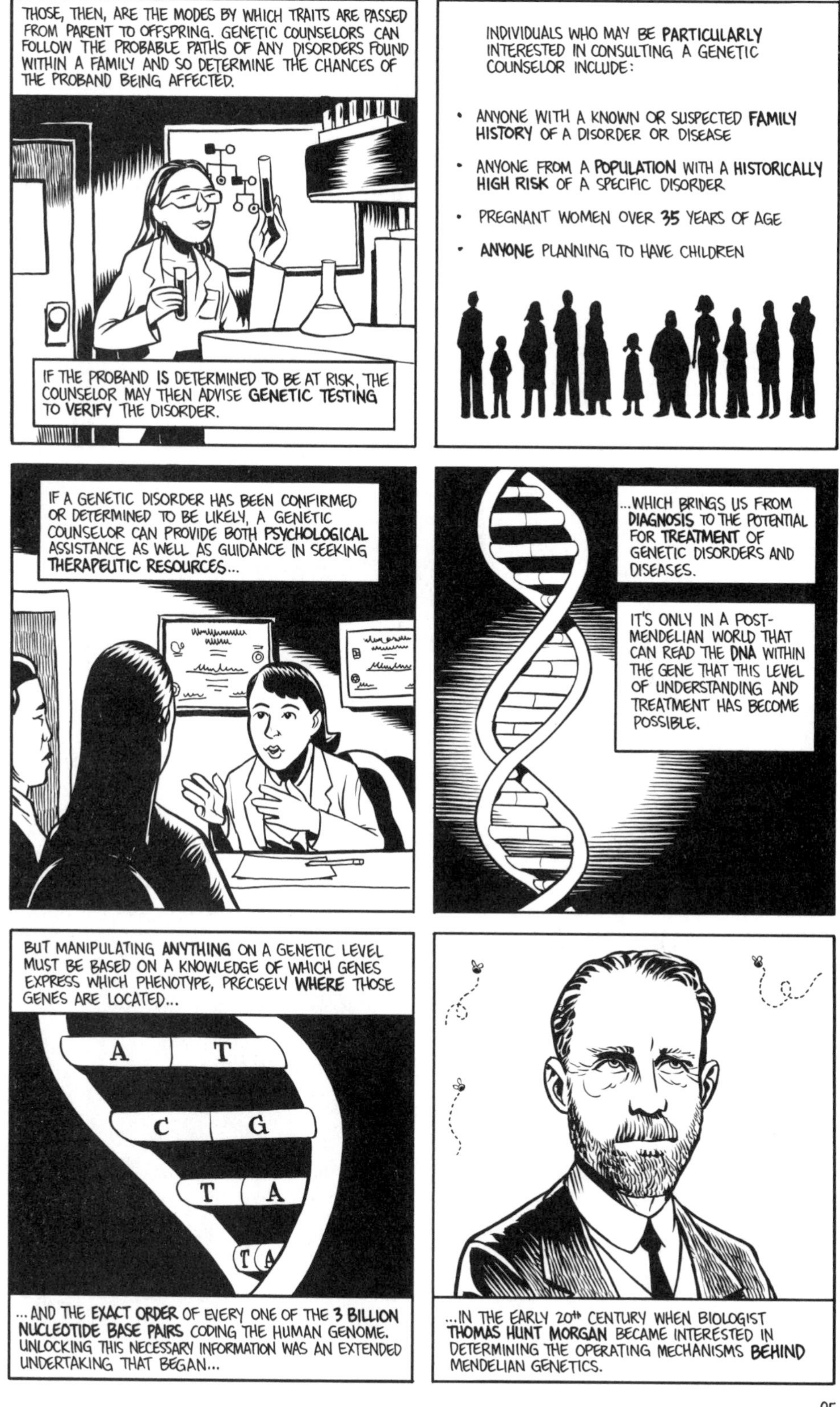
THOSE, THEN, ARE THE MODES BY WHICH TRAITS ARE PASSED FROM PARENT TO OFFSPRING. GENETIC COUNSELORS CAN FOLLOW THE PROBABLE PATHS OF ANY DISORDERS FOUND WITHIN A FAMILY AND SO DETERMINE THE CHANCES OF THE PROBAND BEING AFFECTED.
IF THE PROBAND IS DETERMINED TO BE AT RISK, THE COUNSELOR MAY THEN ADVISE GENETIC TESTING TO VERIFY THE DISORDER.
INDIVIDUALS WHO MAY BE PARTICULARLY INTERESTED IN CONSULTING A GENETIC COUNSELOR INCLUDE:
• ANYONE WITH A KNOWN OR SUSPECTED FAMILY HISTORY OF A DISORDER OR DISEASE
• ANYONE FROM A POPULATION WITH A HISTORICALLY HIGH RISK OF A SPECIFIC DISORDER
• PREGNANT WOMEN OVER 35 YEARS OF AGE
• ANYONE PLANNING TO HAVE CHILDREN
IF A GENETIC DISORDER HAS BEEN CONFIRMED OR DETERMINED TO BE LIKELY, A GENETIC COUNSELOR CAN PROVIDE BOTH PSYCHOLOGICAL ASSISTANCE AS WELL AS GUIDANCE IN SEEKING THERAPEUTIC RESOURCES...
...WHICH BRINGS US FROM DIAGNOSIS TO THE POTENTIAL FOR TREATMENT OF GENETIC DISORDERS AND DISEASES.
IT'S ONLY IN A POST-MENDELIAN WORLD THAT CAN READ THE DNA WITHIN THE GENE THAT THIS LEVEL OF UNDERSTANDING AND TREATMENT HAS BECOME POSSIBLE.
BUT MANIPULATING ANYTHING ON A GENETIC LEVEL MUST BE BASED ON A KNOWLEDGE OF WHICH GENES EXPRESS WHICH PHENOTYPE, PRECISELY WHERE THOSE GENES ARE LOCATED...
A T
C G
T A
T A
...AND THE EXACT ORDER OF EVERY ONE OF THE 3 BILLION NUCLEOTIDE BASE PAIRS CODING THE HUMAN GENOME. UNLOCKING THIS NECESSARY INFORMATION WAS AN EXTENDED UNDERTAKING THAT BEGAN...
...IN THE EARLY 20th CENTURY WHEN BIOLOGIST THOMAS HUNT MORGAN BECAME INTERESTED IN DETERMINING THE OPERATING MECHANISMS BEHIND MENDELIAN GENETICS.

TO THAT END, MORGAN AND HIS STUDENTS BEGAN OBSERVING FRUIT FLIES FOR MUTANT VARIATIONS. THEY CROSSBRED MUTATIONS TO OBSERVE PATTERNS OF INHERITANCE.
THERE'S A GOOD BUZZ SURROUNDING OUR RESEARCH.
THEY WERE ABLE TO IDENTIFY SOME COMPLEX GENE TRANSMISSIONS -- INCLUDING THOSE INVOLVING GENES LOCATED ON THE SEX CHROMOSOMES -- AND RECOGNIZED THAT LINKAGES COULD OCCUR BETWEEN GENES LOCATED CLOSELY TOGETHER.
THIS LED TO THEIR BREAKTHROUGH REALIZATION THAT GENE LOCATIONS COULD BE DETERMINED WITHIN THE CHROMOSOMES!
WE'VE BRED THIS ONE TO BE VERY COMPLEX!
ALFRED STURTEVANT, ONE OF MORGAN'S RESEARCHERS, DISCOVERED THE PRINCIPLES UNDERLYING GENE MAPPING. HE WAS ABLE TO MANAGE AND INTERPRET HIS ACCUMULATING GENETIC DATA BY CREATING SCHEMATIC DRAWINGS POSITIONING FRUIT FLY GENES IN RELATION TO ONE ANOTHER ON THEIR CHROMOSOMES.
AND SO THE FIRST UNDERSTANDING OF THE PHYSICAL ORGANIZATION OF GENES HAD BEGUN--
-- THE FIRST STEPS TOWARD ASSOCIATING SPECIFIC TRAITS WITH SPECIFIC GENES FOUND AT SPECIFIC CHROMOSOMAL LOCATIONS.
FIFTY YEARS LATER, WATSON AND CRICK CRACKED OPEN THE DNA MACROMOLECULES OF WHICH CHROMOSOMES ARE BUILT...
...AND IDENTIFIED THE SEQUENTIAL ARRANGEMENTS OF THE FOUR BASES A, T, C, AND G AS THE KEYS TO READING THE INSTRUCTIONS ENCODED IN GENES...

...OPENING UP A NEW **WORLD** OF POSSIBILITIES. THE **LANGUAGE** CONTROLLING GENE EXPRESSION HAD BEEN DISCOVERED. IT WAS NOW WITHIN THE REALM OF POSSIBILITY THAT HUMANS MIGHT EVENTUALLY BE ABLE TO ALTER THEIR GENETIC FATE.

...SEQUENCED IN THE ENDLESS COMBINATIONS THAT MAKE UP EVERY EARTHLY ORGANISM'S INSTRUCTIONS FOR ASSEMBLY AND MAINTENANCE.

THEN, IN THE MID-1970S, BIOCHEMIST **FREDERICK SANGER** DEVELOPED AN EFFICIENT METHOD FOR **SEQUENCING** DNA -- FOR **DETERMINING** THE BASE-PAIR SEQUENCES IN LONG CHAINS OF DNA -- AND THE DOOR SWUNG WIDE OPEN ON THE FUTURE OF GENETICS!

SANGER'S WORK SET THE STAGE FOR MANY IMPORTANT DEVELOPMENTS, BUT PERHAPS THE MOST IMPORTANT WAS THAT OF A GLOBALLY COOPERATIVE UNDERTAKING OF STAGGERING COMPLEXITY AND ENDLESS POTENTIAL...
BIOLOGY
PHYSICS
ETHICS
INFORMATICS
ENGINEERING
CHEMISTRY
...THE HUMAN GENOME PROJECT!
THE HGP WAS INITIATED IN 1990, WITH THE EXPRESS GOAL OF SEQUENCING ALL 3 BILLION BASE PAIRS IN THE HUMAN GENOME...
...BUT A LOT OF NONHUMAN GENETIC INFORMATION WOULD ALSO BE GATHERED ALONG THE WAY!
THE OBJECTIVE WAS TO PROVIDE RESEARCHERS AN UNPRECEDENTED DEPTH AND BREADTH OF UNDERSTANDING OF THE GENETIC FACTORS BEHIND HUMAN DISEASES AND DISORDERS...
...AND THE POTENTIAL FOR CREATING NEW STRATEGIES OF DIAGNOSIS, PREVENTION, AND TREATMENT.
THE GOAL WAS ACTUALLY ACCOMPLISHED TWO YEARS AHEAD OF SCHEDULE, THANKS IN LARGE PART TO CONTINUAL ADVANCES IN DNA SEQUENCING METHODOLOGY AND RELIABILITY.
IT WAS A TRULY AMAZING ACCOMPLISHMENT...
...BY ANY INTELLIGENT SPECIES' STANDARDS!
IN 2003 THE HGP CELEBRATED PROJECT COMPLETION. THE HUMAN GENOME HAD BEEN SEQUENCED TO A DEGREE OF 99.99% ACCURACY.
H.G.P.
THE INTERNATIONAL STANDARD SETS THE DEGREE OF ACCEPTABLE ERROR AT ONE FOR EVERY 10,000 BASE PAIRS. 99.99% -- THAT'S PRETTY DARN ACCURATE.
BY THE WAY, THE SEQUENCED GENOME WAS NOT THAT OF ANY ONE SPECIFIC HUMAN. IT IS A REFERENTIAL SEQUENCE BASED ON DNA SAMPLED FROM GLOBALLY DISPERSED POPULATIONS.
REMEMBER THAT GENES COME IN ALLELIC VARIATIONS. ANY INDIVIDUAL'S PERSONAL GENOME WILL VARY FROM ANYONE ELSE'S BY ABOUT .01%.

BUT **BEFORE** THE HGP COMPLETED THE **HUMAN** GENOME, THEY DECIPHERED THE SEQUENCING OF SOME MUCH SIMPLER ORGANISMS.
MANY GENES ARE PRACTICALLY **IDENTICAL** FROM SPECIES TO SPECIES-- ESPECIALLY THOSE RESPONSIBLE FOR METABOLIC PROCESSES. LEARNING THE BASE SEQUENCES THAT COMPRISE THE GENES OF **ANY** ORGANISM IS LIKELY TO YIELD PLENTY OF INFORMATION ABOUT **OTHER** ORGANISMS.
YOU AND I, WE HAVE A LOT IN COMMON.
RESEARCHERS DID NOT HAVE TO START FROM **SCRATCH** TO DECIPHER THE BASE SEQUENCES OF THE APPROXIMATELY 25,000 HUMAN GENES. **MANY** HAD BEEN SEQUENCED WHILE WORKING ON THE GENOMES OF OTHER, LESS COMPLEX ORGANISMS.
IT SEEMS WE SHARE A GENE FOR **CHEESE** APPRECIATION.
THE VERY **FIRST** GENOME COMPLETELY SEQUENCED -- IN **1995** -- WAS THAT OF A STRAIN OF **BACTERIA**. IT HAS A RELATIVELY MANAGEABLE COMPLEMENT OF **1.9 MILLION BASE PAIRS** AND **1,700 GENES**.
WE'RE ALSO THE MOST **SUCCESSFUL** ORGANISMS ON EARTH! KEEP IT **SIMPLE**, STUPID!
BREWER'S YEAST WAS THE FIRST **EUKARYOTE** SEQUENCED. THIS FUNGUS IS AN ORGANISM CONSIDERABLY MORE CLOSELY RELATED TO HUMANS, WITH **30%** OF ITS GENES **ANALOGOUS** TO HUMAN GENES.
WE SHARE A COMMON INTEREST, **TOO**...
BREWER'S YEAST IS IMPORTANT IN **FOOD PRODUCTION**-- IT MAKES BREAD **RISE** AND **FERMENTS** BEER AND WINE.
SEQUENCING ITS GENOME PROVIDED VALUABLE INFORMATION TO **AGRICULTURAL SCIENCE**...
hic

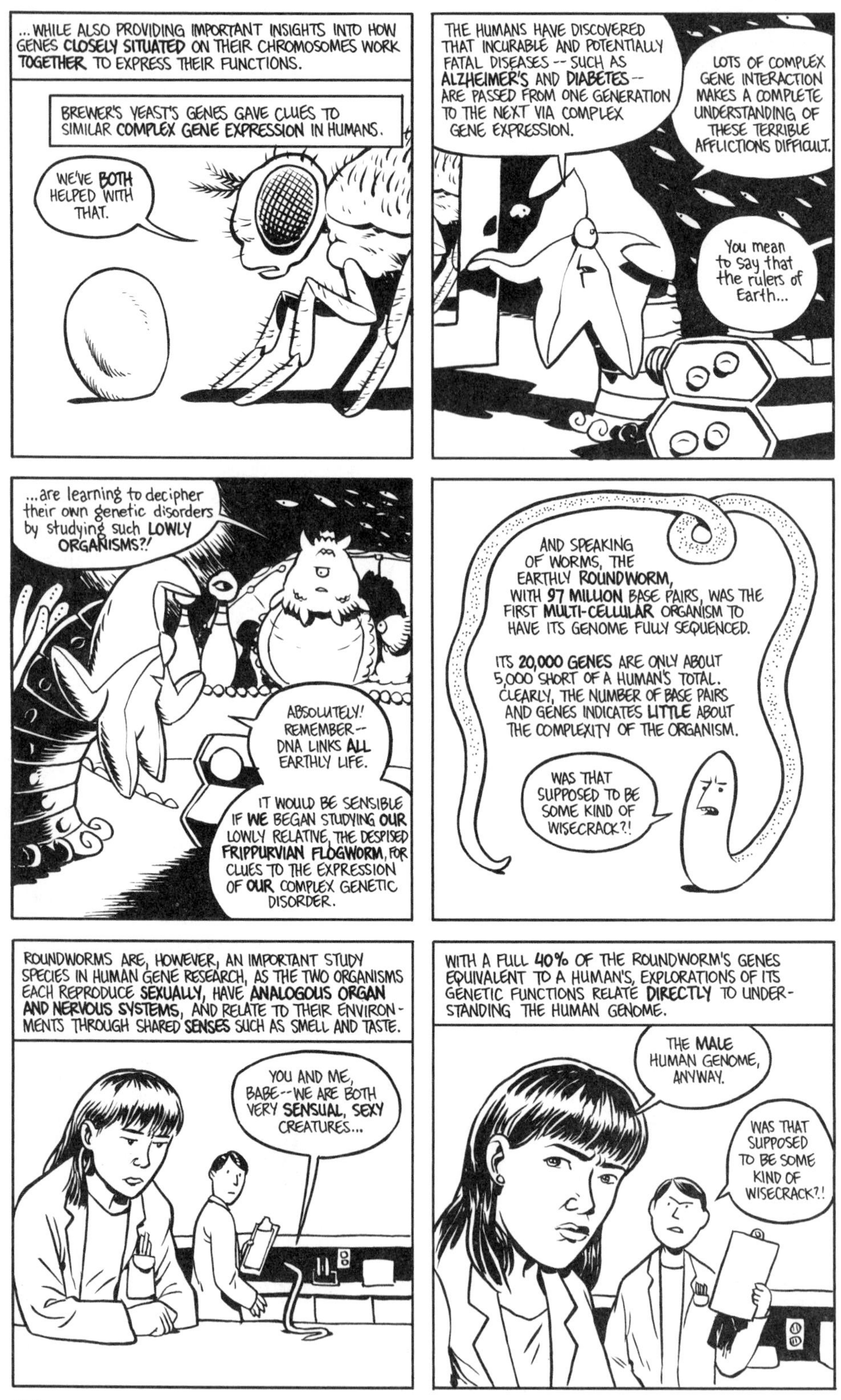

...WHILE ALSO PROVIDING IMPORTANT INSIGHTS INTO HOW GENES CLOSELY SITUATED ON THEIR CHROMOSOMES WORK TOGETHER TO EXPRESS THEIR FUNCTIONS.
BREWER'S YEAST'S GENES GAVE CLUES TO SIMILAR COMPLEX GENE EXPRESSION IN HUMANS.
WE'VE BOTH HELPED WITH THAT.
THE HUMANS HAVE DISCOVERED THAT INCURABLE AND POTENTIALLY FATAL DISEASES -- SUCH AS ALZHEIMER'S AND DIABETES -- ARE PASSED FROM ONE GENERATION TO THE NEXT VIA COMPLEX GENE EXPRESSION.
LOTS OF COMPLEX GENE INTERACTION MAKES A COMPLETE UNDERSTANDING OF THESE TERRIBLE AFFLICTIONS DIFFICULT.
You mean to say that the rulers of Earth...
...are learning to decipher their own genetic disorders by studying such LOWLY ORGANISMS?!
ABSOLUTELY! REMEMBER-- DNA LINKS ALL EARTHLY LIFE.
IT WOULD BE SENSIBLE IF WE BEGAN STUDYING OUR LOWLY RELATIVE, THE DESPISED FRIPPURVIAN FLOGWORM, FOR CLUES TO THE EXPRESSION OF OUR COMPLEX GENETIC DISORDER.
AND SPEAKING OF WORMS, THE EARTHLY ROUNDWORM, WITH 97 MILLION BASE PAIRS, WAS THE FIRST MULTI-CELLULAR ORGANISM TO HAVE ITS GENOME FULLY SEQUENCED.
ITS 20,000 GENES ARE ONLY ABOUT 5,000 SHORT OF A HUMAN'S TOTAL. CLEARLY, THE NUMBER OF BASE PAIRS AND GENES INDICATES LITTLE ABOUT THE COMPLEXITY OF THE ORGANISM.
WAS THAT SUPPOSED TO BE SOME KIND OF WISECRACK?!
ROUNDWORMS ARE, HOWEVER, AN IMPORTANT STUDY SPECIES IN HUMAN GENE RESEARCH, AS THE TWO ORGANISMS EACH REPRODUCE SEXUALLY, HAVE ANALOGOUS ORGAN AND NERVOUS SYSTEMS, AND RELATE TO THEIR ENVIRONMENTS THROUGH SHARED SENSES SUCH AS SMELL AND TASTE.
YOU AND ME, BABE--WE ARE BOTH VERY SENSUAL, SEXY CREATURES...
WITH A FULL 40% OF THE ROUNDWORM'S GENES EQUIVALENT TO A HUMAN'S, EXPLORATIONS OF ITS GENETIC FUNCTIONS RELATE DIRECTLY TO UNDERSTANDING THE HUMAN GENOME.
THE MALE HUMAN GENOME, ANYWAY.
WAS THAT SUPPOSED TO BE SOME KIND OF WISECRACK?!

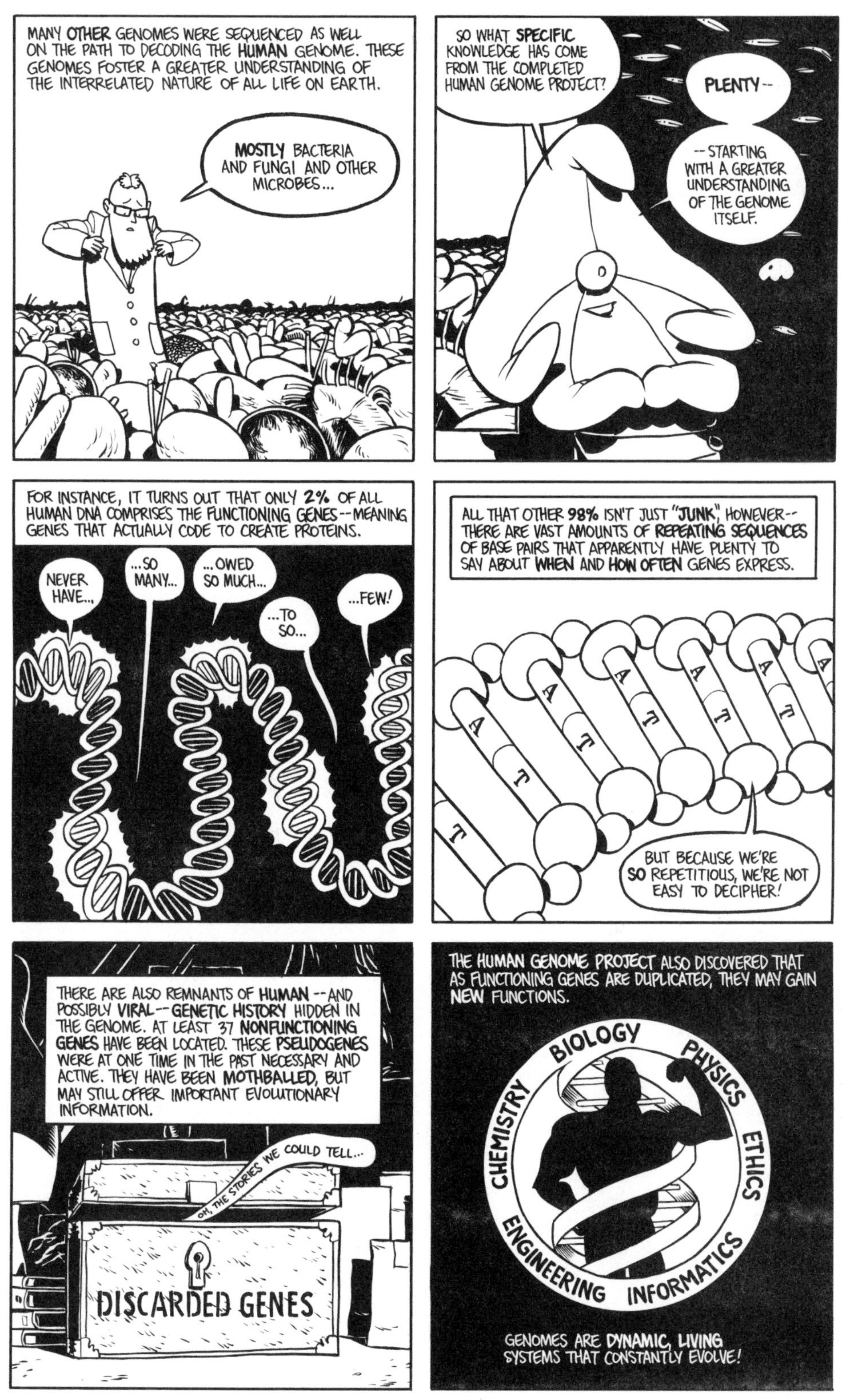
MANY **OTHER** GENOMES WERE SEQUENCED AS WELL ON THE PATH TO DECODING THE **HUMAN** GENOME. THESE GENOMES FOSTER A GREATER UNDERSTANDING OF THE INTERRELATED NATURE OF ALL LIFE ON EARTH.
MOSTLY BACTERIA AND FUNGI AND OTHER MICROBES...
SO WHAT **SPECIFIC** KNOWLEDGE HAS COME FROM THE COMPLETED HUMAN GENOME PROJECT?
PLENTY--
--STARTING WITH A GREATER UNDERSTANDING OF THE GENOME ITSELF.
FOR INSTANCE, IT TURNS OUT THAT ONLY **2%** OF ALL HUMAN DNA COMPRISES THE **FUNCTIONING GENES**--MEANING GENES THAT ACTUALLY CODE TO CREATE PROTEINS.
NEVER HAVE...
...SO MANY...
...OWED SO MUCH...
...TO SO...
...FEW!
ALL THAT OTHER **98%** ISN'T JUST "**JUNK**," HOWEVER--THERE ARE VAST AMOUNTS OF **REPEATING SEQUENCES** OF BASE PAIRS THAT APPARENTLY HAVE PLENTY TO SAY ABOUT **WHEN** AND **HOW OFTEN** GENES EXPRESS.
T A T A T A T A T A
BUT BECAUSE WE'RE **SO** REPETITIOUS, WE'RE NOT EASY TO DECIPHER!
THERE ARE ALSO REMNANTS OF **HUMAN**--AND POSSIBLY **VIRAL**--**GENETIC HISTORY** HIDDEN IN THE GENOME. AT LEAST 37 **NONFUNCTIONING GENES** HAVE BEEN LOCATED. THESE **PSEUDOGENES** WERE AT ONE TIME IN THE PAST NECESSARY AND ACTIVE. THEY HAVE BEEN **MOTHBALLED**, BUT MAY STILL OFFER IMPORTANT EVOLUTIONARY INFORMATION.
OH, THE STORIES WE COULD TELL...
DISCARDED GENES
THE **HUMAN GENOME PROJECT** ALSO DISCOVERED THAT AS FUNCTIONING GENES ARE DUPLICATED, THEY MAY GAIN **NEW** FUNCTIONS.
BIOLOGY
PHYSICS
ETHICS
INFORMATICS
ENGINEERING
CHEMISTRY
GENOMES ARE **DYNAMIC, LIVING** SYSTEMS THAT CONSTANTLY EVOLVE!

And I suppose that this DYNAMIC ACTIVITY is due to their SEXUAL NATURE and all very GOOD for the species' health...
WELL, IT'S DUE TO THE GENETIC VARIETY AVAILABLE THROUGH SEXUAL REPRODUCTION PLUS THE VARIATIONS THAT OCCUR THROUGH NORMAL MUTATION.
AND WHILE GENOME CHANGES ARE NEGLIGIBLE FROM A GENERATIONAL PERSPECTIVE, OVER TIME THEY KEEP SPECIES COMPETITIVE AND VITAL.
And THAT'S what we SQUINCH seem to be missing.
WE CERTAINLY COULD USE SOME FRESH GENETIC ALTERNATIVES TO MEET THE DISEASES THAT ARE OVER-TAKING US.
BUT, LET ME GET BACK TO THE HUMAN GENOME...
THE COMPLETED HUMAN GENOME ANNOUNCED IN 2003 WAS ACTUALLY A ROUGH DRAFT OF THE BASE-PAIR TEXT FOUND ON EACH CHROMOSOME.
THERE WAS STILL MUCH WORK TO BE DONE -- THAT 3-BILLION-BASE-PAIR-LONG CHAIN OF NUCLEOTIDES NEEDED TO BE INTERPRETED AS SPECIFIC GENES, REPEATING SEQUENCES, AND EVERYTHING ELSE THAT IS HELD IN THE GENOME.
?
1 2 3 4 5 6 7 8 9 10 11 12 13 14 15 16 17 18 19 20 21 22 X/Y
FURTHER ADVANCES IN HIGH-QUALITY SEQUENCING TECHNIQUES HAVE HELPED FILL IN MUCH OF THE UNKNOWN TERRITORY ON THE GENETIC MAP, AND HAVE PROVIDED ACCURATE LOCATIONS FOR THE GENES ON ALL 24 HUMAN CHROMOSOMES.
FOR INSTANCE, THE MAP OF CHROMOSOME 1 -- DESIGNATED AS SUCH BECAUSE IT IS THE BIGGEST HUMAN CHROMOSOME -- PLACES 3,141 GENES AND 991 PSEUDOGENES.
OF PARTICULAR MEDICAL INTEREST WAS THE DISCOVERY THAT MUTATIONS AFFECTING SEVERAL GENES ON CHROMOSOME 1 HAVE A DIRECT CORRELATION TO THE APPEARANCE OF CERTAIN CANCERS.

BUT THE GENES MATCHED WITH SPECIFIC FUNCTIONS--OR EVEN DISORDERS--REPRESENT JUST A FRACTION OF ALL GENES.
= ONE GENE CONTRIBUTING TO EYE COLOR (BLUE)
THERE ARE STILL HUGE SWATHS OF THE HUMAN GENOME THAT NEED TO BE INTERPRETED...
...AND HUMAN SCIENTISTS ARE WORKING TO MAKE WHAT IS KNOWN OF THE GENOME MORE USER-FRIENDLY. FOR EXAMPLE...
...CANCER RESEARCHERS, IN 2006, BEGAN A MONUMENTAL FOLLOW-UP TO THE HUMAN GENOME PROJECT CALLED...
THE CANCER GENOME ATLAS
...THE CANCER GENOME ATLAS!
CANCERS ARE ABNORMAL, UNCONTROLLED CELL GROWTHS THAT EXACT A TERRIBLE TOLL ON HUMANKIND.
EVEN IN EARTH'S MOST TECHNOLOGICALLY ADVANCED NATION, THE U.S.A., ONE PERSON DIES FROM CANCER EVERY MINUTE.
STARTING WITH THE VAST AMOUNT OF RAW INFORMATION PROVIDED BY THE HGP, THE GOAL OF THE ATLAS IS TO EVENTUALLY IDENTIFY AND INDEX ALL GENES LINKED TO CANCERS...
...PROVIDING A POWERFUL, SPECIALIZED TOOL FOR ONGOING RESEARCH AND THERAPY DEVELOPMENT.
THAT WOULD BE A HELP...
THE HUMAN GENOME
LINKING THOSE GENES TO THEIR FUNCTIONS HAS ALREADY LED TO SOME IMPORTANT DISCOVERIES.
THE GENE DESIGNATED BRCA 1, FOR INSTANCE...

BRCA 1 HAS BEEN DEMONSTRATED TO CODE FOR A PROTEIN THAT NORMALLY ORCHESTRATES THE REPAIR OF DAMAGED DNA. BUT IT ALSO APPEARS TO MODERATE THE EFFECT OF THE GROWTH HORMONE PROGESTERONE...
GET THAT G OUT AND THAT C IN-- AND BE QUICK ABOUT IT!
BRCA 1
...ON A WOMAN'S BREAST AND OVARY CELLS. IT HELPS DESTROY THE PROTEIN RECEPTORS THAT ALLOW THOSE CELLS TO RECEIVE PROGESTERONE'S MESSAGE...
...ONCE PROPER GROWTH HAS BEEN ATTAINED AND BEFORE THE HORMONE CAN URGE TOO MUCH UNCONTROLLED GROWTH.
PROGESTERONE
RECEPTOR
OVARY CELL
BRCA 1
THAT'S ENOUGH, ALREADY! GET OUTTA HERE!
PROGESTERONE'S MESSAGE
MUTATIONS TO BRCA 1, HOWEVER, DISRUPT THIS PROCESS, LEAVING THE GENE'S ROLE IN SUPERVISING NORMAL CELL GROWTH INCOMPLETE AND SO OPENING THE DOOR TO UNREGULATED AND INVASIVE CELL GROWTH...
...A HALLMARK OF CANCER.
BRCA 1
CAN'T REMEMBER WHAT TO DO NEXT...
BRCA 1
MUTATED BRCA 1 GENES ARE JUST ONE FACTOR IN THE DEVELOPMENT OF BREAST AND OVARIAN CANCER. BEING ABLE TO TARGET BRCA 1 AS WELL AS ALL OTHER CULPRITS HOLDS THE PROMISE OF DEVELOPING THERAPIES THAT MIGHT CORRECT HARMFUL MUTATIONS AND SO BE MORE EFFECTIVE, AND LESS TRAUMATIC, THAN CURRENT SURGICAL, RADIATION, AND CHEMOTHERAPY TREATMENTS.
GOING FORWARD, CANCER GENOME ATLAS RESEARCHERS WILL COMPARE THE DNA OF TUMOR SAMPLES AGAINST THE HGP'S REFERENTIAL HUMAN GENOME, EXAMINING GENES SUSPECTED OF ASSOCIATION WITH SPECIFIC CANCERS AND SEARCHING FOR ABNORMALITIES.
The Human GENOME from A to T!
THE CANCER GENOME ATLAS
AND ALTHOUGH PREVAILING THEORY HAS LONG HAD IT THAT MUTATIONS IN KEY REGULATORY GENES ARE THE PRIMARY CAUSE OF CANCER, IT HAS RECENTLY BEEN SUGGESTED THAT ABNORMALITIES ON A CHROMOSOMAL LEVEL MAY BE AS MUCH TO BLAME!
WHAT?!

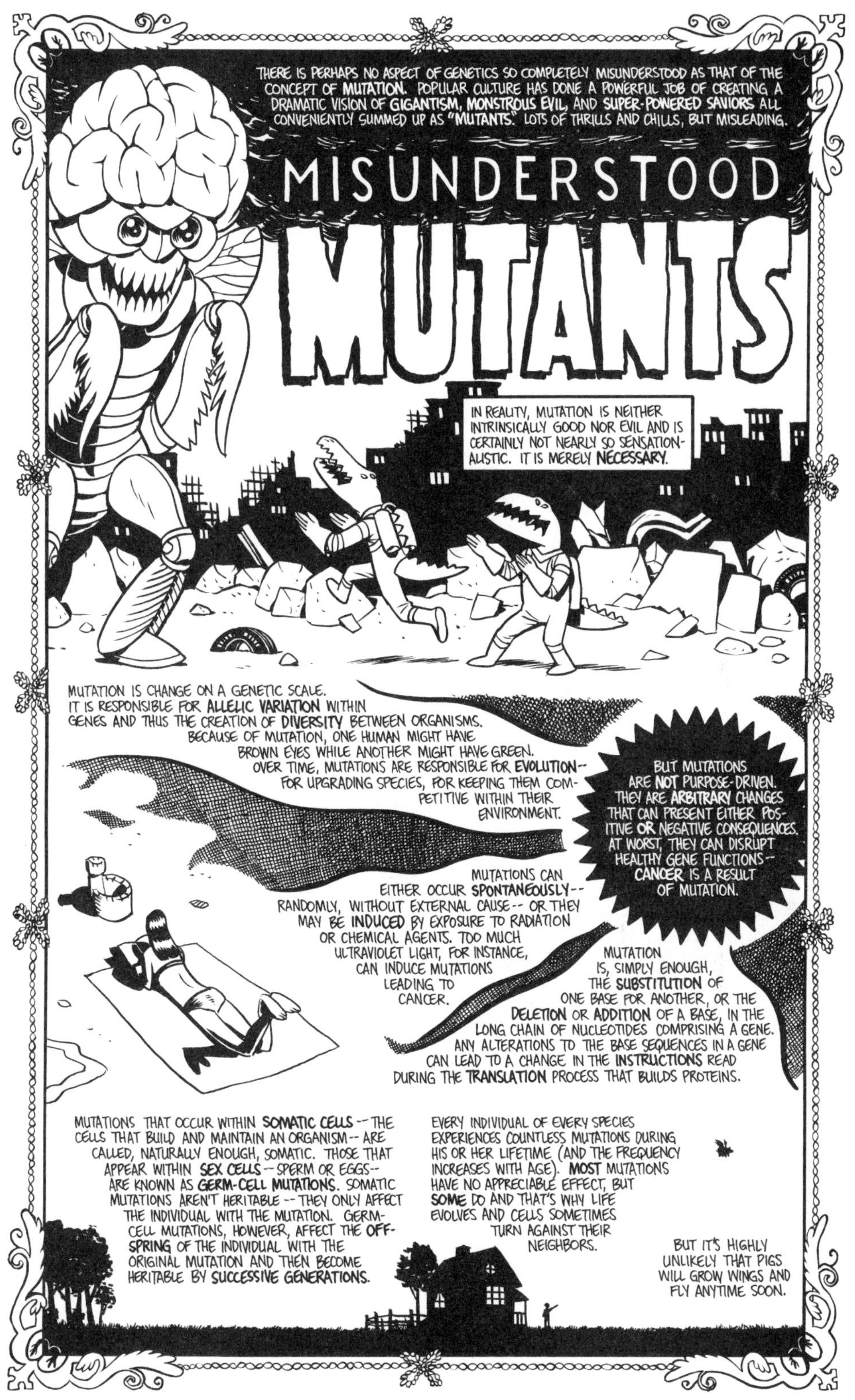

THERE IS PERHAPS NO ASPECT OF GENETICS SO COMPLETELY MISUNDERSTOOD AS THAT OF THE CONCEPT OF **MUTATION**. POPULAR CULTURE HAS DONE A POWERFUL JOB OF CREATING A DRAMATIC VISION OF **GIGANTISM, MONSTROUS EVIL**, AND **SUPER-POWERED SAVIORS** ALL CONVENIENTLY SUMMED UP AS **"MUTANTS."** LOTS OF THRILLS AND CHILLS, BUT MISLEADING.
MISUNDERSTOOD
MUTANTS
IN REALITY, MUTATION IS NEITHER INTRINSICALLY GOOD NOR EVIL AND IS CERTAINLY NOT NEARLY SO SENSATIONALISTIC. IT IS MERELY **NECESSARY**.
MUTATION IS CHANGE ON A GENETIC SCALE. IT IS RESPONSIBLE FOR **ALLELIC VARIATION** WITHIN GENES AND THUS THE CREATION OF **DIVERSITY** BETWEEN ORGANISMS. BECAUSE OF MUTATION, ONE HUMAN MIGHT HAVE BROWN EYES WHILE ANOTHER MIGHT HAVE GREEN. OVER TIME, MUTATIONS ARE RESPONSIBLE FOR **EVOLUTION**-- FOR UPGRADING SPECIES, FOR KEEPING THEM COMPETITIVE WITHIN THEIR ENVIRONMENT.
BUT MUTATIONS ARE **NOT** PURPOSE-DRIVEN. THEY ARE **ARBITRARY** CHANGES THAT CAN PRESENT EITHER POSITIVE **OR** NEGATIVE CONSEQUENCES. AT WORST, THEY CAN DISRUPT HEALTHY GENE FUNCTIONS-- **CANCER** IS A RESULT OF MUTATION.
MUTATIONS CAN EITHER OCCUR **SPONTANEOUSLY**-- RANDOMLY, WITHOUT EXTERNAL CAUSE-- OR THEY MAY BE **INDUCED** BY EXPOSURE TO RADIATION OR CHEMICAL AGENTS. TOO MUCH ULTRAVIOLET LIGHT, FOR INSTANCE, CAN INDUCE MUTATIONS LEADING TO CANCER.
MUTATION IS, SIMPLY ENOUGH, THE **SUBSTITUTION** OF ONE BASE FOR ANOTHER, OR THE **DELETION** OR **ADDITION** OF A BASE, IN THE LONG CHAIN OF NUCLEOTIDES COMPRISING A GENE. ANY ALTERATIONS TO THE BASE SEQUENCES IN A GENE CAN LEAD TO A CHANGE IN THE **INSTRUCTIONS** READ DURING THE **TRANSLATION** PROCESS THAT BUILDS PROTEINS.
MUTATIONS THAT OCCUR WITHIN **SOMATIC CELLS**-- THE CELLS THAT BUILD AND MAINTAIN AN ORGANISM-- ARE CALLED, NATURALLY ENOUGH, SOMATIC. THOSE THAT APPEAR WITHIN **SEX CELLS**-- SPERM OR EGGS-- ARE KNOWN AS **GERM-CELL MUTATIONS**. SOMATIC MUTATIONS AREN'T HERITABLE-- THEY ONLY AFFECT THE INDIVIDUAL WITH THE MUTATION. GERM-CELL MUTATIONS, HOWEVER, AFFECT THE **OFFSPRING** OF THE INDIVIDUAL WITH THE ORIGINAL MUTATION AND THEN BECOME HERITABLE BY **SUCCESSIVE GENERATIONS**.
EVERY INDIVIDUAL OF EVERY SPECIES EXPERIENCES COUNTLESS MUTATIONS DURING HIS OR HER LIFETIME (AND THE FREQUENCY INCREASES WITH AGE). **MOST** MUTATIONS HAVE NO APPRECIABLE EFFECT, BUT **SOME** DO AND THAT'S WHY LIFE EVOLVES AND CELLS SOMETIMES TURN AGAINST THEIR NEIGHBORS.
BUT IT'S HIGHLY UNLIKELY THAT PIGS WILL GROW WINGS AND FLY ANYTIME SOON.

KARYOTYPE OF NORMAL HUMAN CHROMOSOME SETS
1 2 3 4 5 6 7 8 9 10 11 12 13 14 15 16 17 18 19 20 21 22 X/Y
MANY GENETIC DISORDERS ARE INDISPUTABLY THE RESULT OF CHROMOSOMAL ABNORMALITIES--ANEUPLOIDY IS A DISORDER IN WHICH THERE IS AN IMBALANCE IN THE NUMBER OF CHROMOSOMES.
REMEMBER--HUMANS ARE NORMALLY DIPLOID, MEANING THAT THEY CARRY TWO SETS OF THEIR 22 AUTOSOMAL CHROMOSOMES, PLUS TWO SEX CHROMOSOMES (EITHER XX OR XY).
ANEUPLOIDY OCCURS WHEN CHROMOSOMES DO NOT SEGREGATE PROPERLY DURING MEIOSIS, LEAVING EGG OR SPERM CELLS WITH EITHER TOO MANY OR TOO FEW CHROMOSOMES.
CHROMOSOMAL IMBALANCE IS RESPONSIBLE FOR 50% OF ALL CASES OF MISCARRIAGES...
NONDISJUNCTION
FIRST MEIOTIC DIVISION
SECOND MEIOTIC DIVISION
FERTILIZATION
TRISOMIC
(ONE TOO MANY CHROMOSOMES)
MONOSOMIC
(ONE TOO FEW CHROMOSOMES)
DOWN SYNDROME-- EXTRA CHROMOSOME 21
...AND THOSE OFFSPRING THAT DO COME TO TERM ARE AFFLICTED WITH BIRTH DEFECTS. DOWN, CRI DU CHAT, AND KLINEFELTER'S SYNDROMES ARE AMONG THE MANY POSSIBLE DISORDERS.
RECENT ADVANCES IN DIAGNOSTICS, AIMED AT DETERMINING THE STATE OF PLOIDY BEFORE PREGNANCY, OFFER THE POSSIBILITY OF GUIDANCE UNAVAILABLE TO EARLIER GENERATIONS.
AND, AS MENTIONED BEFORE, ANEUPLOIDY IS COMMONLY SEEN WITHIN CANCER CELLS.
SO, RESEARCH INTO ALL THESE ABNORMALITIES CONTINUES ON MANY FRONTS...
Okay, okay!
I UNDERSTAND-- humans are getting a much better grip on IDENTIFYING the genetic sources of many of their maladies...
...but I'm still waiting to hear what they're doing to CORRECT their problems!
I WAS JUST GETTING THERE, MY MOST IMPATIENT POTENTATE!

HUMAN SCIENTISTS HAVE BEGUN TO WORK WITHIN THE GENE, MANIPULATING DNA AND EVEN INTRODUCING IT TO NEW LOCATIONS. IT'S ALL ABOUT EXPRESSING THE RIGHT PROTEIN IN THE RIGHT LOCATION AT THE RIGHT TIME.
WELCOME TO THE WILD NEW WORLD OF GENETIC ENGINEERING!
IT ALL BEGAN IN 1970, WHEN MICROBIOLOGIST HAMILTON O. SMITH IDENTIFIED CERTAIN RESTRICTION ENZYMES WITH WHICH HE WAS ABLE TO SEPARATE NUCLEOTIDES WITHIN A STRAND OF DNA.
EXPERIMENTING FURTHER, HE DISCOVERED THAT WITH ANOTHER ENZYME -- LIGASE -- HE COULD PASTE THOSE DISASSEMBLED BITS OF DNA INTO NEW ARRANGEMENTS.
THIS TECHNIQUE IS CALLED GENE SPLICING...
LIGASE
...OR RECOMBINANT DNA TECHNOLOGY.
DNA IS CUT APART AND SPLICED BACK TOGETHER -- RECOMBINED -- INTO NEW BASE-PAIR SEQUENCES IN ORDER TO CREATE ARTIFICIAL GENES.
I FEEL LIKE EXPRESSING MYSELF IN WAYS I NEVER THOUGHT POSSIBLE!
TO WHAT END, YOU MIGHT ASK?
MAYBE MOST IMPORTANT, TO PRODUCE PREVIOUSLY IMPOSSIBLE QUANTITIES OF QUALITY-CONTROLLED PROTEINS FOR THERAPEUTIC SERVICES.
A MAKE-BELIEVE, GROSSLY SIMPLIFIED EXAMPLE: LET'S SAY THAT THERE IS A GROUP OF HUMANS WHO SUFFER FROM THE LACK OF A PARTICULAR PROTEIN. THEIR BODIES CAN'T PRODUCE THAT PROTEIN.
WE WERE BORN WITHOUT THE ABILITY TO SMILE.
THESE HUMANS NEED THIS PROTEIN PROVIDED FOR THEM AFFORDABLY AND IN SUFFICIENT QUANTITY.
THE GENE THAT PRODUCES THE NEEDED PROTEIN IS, OF COURSE, AVAILABLE IN OTHER HUMANS. BUT IN ORDER TO CREATE SUFFICIENT QUANTITIES OF THAT PROTEIN, THE GENE NEEDS TO BE PLACED IN A UNIQUE ENVIRONMENT -- A VECTOR -- THAT WILL ALLOW IT TO REPLICATE RAPIDLY.
I MAKE THE SMILE PROTEIN JUST FINE -- BUT NOT ENOUGH FOR ALL OF US.

THE PERFECT VECTOR JUST HAPPENS TO LIVE CLOSE BY, IN THE HUMAN GUT. IT'S AN ANCIENT BACTERIAL ALLY NAMED ESCHERICHIA COLI.
E. COLI IS THE PERFECT INCUBATOR FOR REPLICATING GENES QUICKLY.
I'VE ALWAYS BEEN THERE FOR YOU!
E. COLI
E. COLI IS A PROKARYOTE WITH A SIMPLE RING OF DNA CALLED A PLASMID, WHICH IS FIRST REMOVED FROM THE BACTERIA.
USING THE RESTRICTION ENZYMES, THE PLASMID IS CUT UP AND ITS UNWANTED GENES ARE REMOVED. THEN OUR HUMAN SMILE GENE IS SPLICED INTO THE RING!
OUT WITH THE BAD...
...AND IN WITH THE GOOD!
LIGASE
PLASMID
THE MODIFIED PLASMID IS PLACED BACK IN ITS BACTERIAL HOME AND THE GENETICALLY MANIPULATED E. COLI IS CULTURED AND ENCOURAGED TO MULTIPLY...
...AND BACTERIA ARE VERY GOOD AT MULTIPLYING.
E. COLI REPLICATES ITSELF TO FORM BILLIONS OF COPIES IN SHORT ORDER. EVERY COPY OF THE MODIFIED PLASMID WILL CARRY A COPY OF THE INSERTED HUMAN GENE...
...WITH THE MUCH-NEEDED SMILE PROTEIN BEING PRODUCED FROM THE DNA INSTRUCTIONS OF EVERY ONE OF THOSE BILLIONS AND BILLIONS OF GENES!
WHY DO I FEEL THE URGE TO SMILE? I DON'T HAVE A MOUTH...
RECOMBINANT DNA TECHNOLOGY HAS ALREADY BEEN USED TO MAKE QUANTITIES OF THESE REAL THERAPEUTIC PROTEINS:
INTERFERON
A POWERFUL VIRUS FIGHTER WITH GREAT PROMISE. USED TO COMBAT CERTAIN CANCERS.
INSULIN
CRITICAL FOR BREAKING DOWN SUGARS IN HUMAN BLOOD. LACKING IN INDIVIDUALS WITH DIABETES.
SOMATOTROPIN
ALSO KNOWN AS HUMAN GROWTH HORMONE. HELPS PREVENT GROWTH DEFICIENCIES, ESPECIALLY IN CHILDREN. AT THIS TIME, THE RECOMBINANT FORM IS STILL BEING TESTED FOR SAFETY.
BUT A SMILE SERUM WOULDN'T BE A BAD IDEA, EITHER!

SMILE
SMILE
THE PROSPECT OF VATS AND VATS OF INEXPENSIVE THERAPEUTIC PROTEINS IS JUST ONE OF GENETIC ENGINEERING'S POTENTIALS.
THE ABILITY TO INSERT THERAPEUTIC GENES DIRECTLY INTO A PATIENT'S AILING CELLS IS ANOTHER...
...MADE POSSIBLE BY THE USE OF VIRUSES AS VECTORS!
VIRUSES ARE BIZARRE THINGS. THEY ARE NEITHER DEAD NOR ALIVE, BUT THEY CARRY THEIR OWN DNA AND THEY REPRODUCE--WHICH MAKES THEM VALUABLE DELIVERY SYSTEMS.
INFLUENZA VIRUS
DNA--OR RNA--IS REALLY THE ONLY THING THAT A VIRUS HAS WITHIN ITS SHELL. TO REPRODUCE, IT MUST CONVERT ITS RNA TO A FORM OF DNA, THEN INVADE A HOST CELL AND HIJACK THAT CELL'S REPLICATION EQUIPMENT.
GIMME ALL YOUR RIBOSOMES AND DON'T ASK QUESTIONS!
SOME VIRUSES USE THE CELL TO KEEP REPLICATING THEMSELVES UNTIL THE CELL IS BLOWN APART BY THE POPULATION BOOM! NOT ALL ARE SO DESTRUCTIVE TO THEIR HOST, HOWEVER.
THE MOST EFFECTIVE VIRUSES WANT TO KEEP THEIR HOST ALIVE!
NOW, THERE IS A PARTICULAR TYPE OF VIRUS CALLED A RETROVIRUS THAT REPRODUCES BY SPLICING ITS NEWLY SYNTHESIZED DNA RIGHT ONTO A HOST CELL'S CHROMOSOMES.
SO, AS THE CHROMOSOMES ARE REPLICATED DURING THE NORMAL CELL CYCLE, SO IS THE VIRAL DNA! IT HAS BECOME A PART OF ITS HOST'S DNA!
GENE SPLICING IS NOTHING NEW FOR US!
THE HUMAN IMMUNODEFICIENCY VIRUS THAT CAUSES ACQUIRED IMMUNE DEFICIENCY SYNDROME (AIDS) WORKS THIS WAY -- AND THAT STRATEGY HAS SO FAR KEPT IT INCURABLE.
BUT, ON THE PLUS SIDE, RETROVIRUSES CAN BE MADE TO FIGHT FOR HUMAN HEALTH!
WE CAN?
HIV RETROVIRUS

VIRUSES, BASED ON THEIR REPRODUCTIVE STRATEGY, ARE THE PERFECT VECTORS FOR WHAT IS CALLED **GENE THERAPY**-- TREATMENTS THAT EITHER INTRODUCE A **NEW, HEALTHY GENE** INTO A COMPROMISED SYSTEM OR **TURN OFF** GENES RESPONSIBLE FOR UNHEALTHY CELL GROWTH.

WITH THE SAME RECOMBINANT TECHNOLOGY USED TO INSERT GENES INTO BACTERIA, SCIENTISTS HAVE RE-OUTFITTED THESE VIRUSES AS DELIVERY SYSTEMS THAT BEAR A RECOMBINED GENETIC **PAYLOAD** STRAIGHT INTO HUMAN CELLS, WHERE THEY THEN DO WHAT COMES NATURALLY-- **REPRODUCE**.

ONCORETROVIRAL VECTORS, DEVELOPED FROM A VIRUS THAT CAUSES LEUKEMIA IN MONKEYS, SPECIALIZE IN REGULATING **ONCOGENES**-- GENES THAT START AN UNCONTROLLED GROWTH PROCESS THAT CAN LEAD TO CANCER.

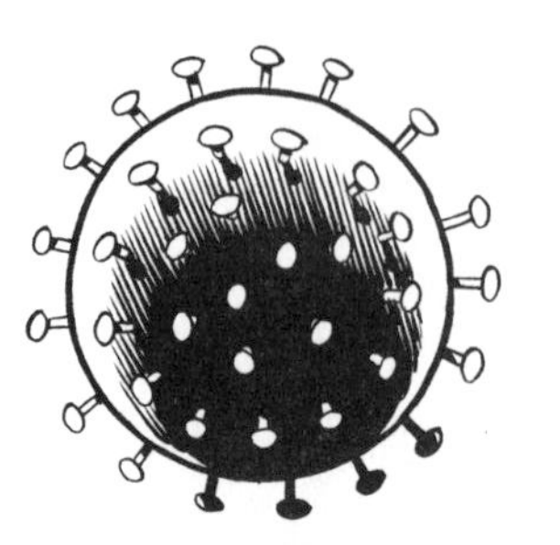

LENTIVIRAL VECTORS ARE DEVELOPED STRAIGHT FROM THE HIV AND CAN DELIVER SIGNIFICANT QUANTITIES OF GENETIC INFORMATION, EVEN SPREADING THEIR ALTERED DNA TO NEIGHBORING CELLS THEY HAVE NOT BODILY INVADED. THEY ARE USED TO PROTECT THE IMMUNE SYSTEM AND ARE VALUABLE TOOLS FOR FIGHTING AIDS.

ADENOVIRAL VECTORS, ON THE OTHER HAND, ARE **NOT** RETROVIRUSES -- THEY DO **NOT** INSERT THEIR DNA INTO THEIR HOST'S DNA. THEY EXIST INDEPENDENTLY WITHIN THE CELL, REPRODUCING REGARDLESS OF CELL DIVISION. THEY CAN RELIABLY DELIVER GENES JUST WHERE THEY ARE NEEDED.

NONE OF THESE GENE-THERAPY OPTIONS IS CLOSE TO PERFECT YET. **ONCORETROVIRAL VECTORS**, WHICH PERFORM WELL ONLY IF CELLS ARE DIVIDING, CAN POTENTIALLY CAUSE CANCER THEMSELVES.

LENTIVIRAL VECTORS CAN CAUSE UNWANTED LOSS-OF-FUNCTION MUTATIONS AND CAN REGAIN THEIR DELETED AND DANGEROUS HIV GENES IF THEY COME IN CONTACT WITH UNALTERED HIV VIRUSES.

ADENOVIRAL VECTORS CAN PRODUCE A STRONG REACTION IN A HOST'S IMMUNE SYSTEM AND ARE NOT DEPENDABLY REPLICATED AND PASSED TO DAUGHTER CELLS.

CURRENT GENETIC THERAPIES CAN TARGET **SPECIFIC CELLS**, AND THERE HAVE EVEN BEEN SUCCESSES ADDRESSING AFFECTED **NUCLEOTIDE SEQUENCES**, ALLOWING THERAPEUTIC GENES TO BE PLACED **ONLY** WHERE THEY ARE REQUIRED. BUT THERE IS A LONG WAY TO GO BEFORE VIRAL GENE THERAPY'S PROMISE IS FULLY REALIZED.

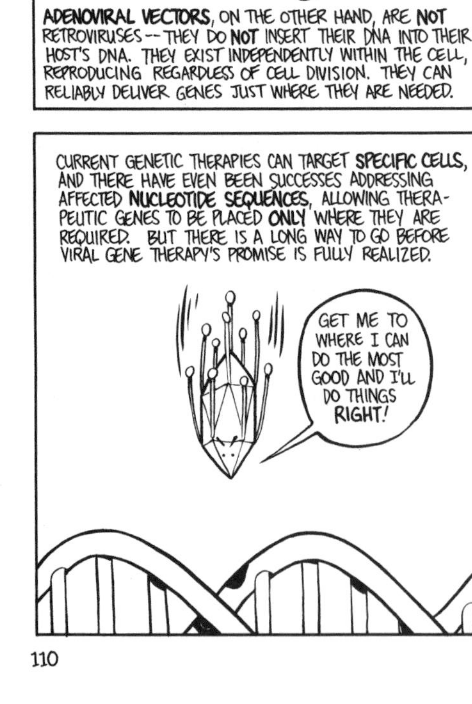

TO THIS END, **DNA LIBRARIES**, WHICH WILL IDENTIFY AND CATALOG ALL ACTIVE GENES IN A SPECIFIC CELL, ARE BEING DEVELOPED TO MAKE GENE SEARCHES EASIER...

...SINCE THE ALTERNATIVE REQUIRES SEARCHING THE **ENTIRE** HUMAN GENOME FOR EACH SPECIFIC GENE.

A DNA LIBRARY WILL GUIDE A RESEARCHER DIRECTLY TO THE GENES KNOWN TO EXPRESS IN ONLY THE CELLS CONCERNED. THOSE AFFLICTED BY--
OKAY, Bloort-- let's hold up again...
CELL G
CELL GENE
...Because, if I'm understanding this gene-therapy business CORRECTLY, it might have some real bearing on our own genetic problems.
Let me try to recap...
...Humans have learned to IDENTIFY the source of many disorders and diseases within their genes, and then to LOCATE the responsible --or contributing-- genes precisely within their chromosomes.
Knowing where to find specific genes, they've learned to MANIPULATE them, first SPLICING together bits of DNA into new sequences...
...and THEN splicing their own genes within the DNA of MICROBES that have the ability to INSERT themselves within human cells.
Humans use these microbial VECTORS to carry therapeutic human genes into afflicted cells, where the inserted DNA can produce PROTEINS to combat the disease!
It's amazing! It's GENIUS!
Shouldn't WE be able to create a similar therapy to combat the mutating disease that's attacking US?!
THAT, I THINK, IS A POSSIBILITY WE CERTAINLY MUST EXPLORE, O MOST PERCEPTIVE POTENTATE.
BUT-- THERE IS EVEN MORE TO CONSIDER...

CLOSELY RELATED TO RECOMBINANT DNA TECHNOLOGY IS TRANSGENIC TECHNOLOGY, WHICH IS USED TO PRODUCE GMOs--
--GENETICALLY MODIFIED ORGANISMS!
TRANSGENIC TECHNOLOGY INVOLVES LIFTING THE GENES FROM ONE ORGANISM AND PLACING THEM WITHIN THE DNA OF ANOTHER ORGANISM.
THIS CAN EVEN INCLUDE PLACING ANIMAL GENES IN PLANTS!
HA! I HEAR YA'.
NOT SURPRISINGLY, THE PUBLIC PERCEPTION OF GMOs IS THAT OF MONSTERS RUNNING WILD. WE'LL SEE THAT WHILE THERE MAY BE A SMIDGEN OF REALITY IN THOSE FEARS...
BOO.
...THE GOALS ARE MORE IN LINE WITH IMPROVED MEDICINAL AND FOOD RESOURCES, AND, OF COURSE, FINANCIAL PROFIT.
AND REMEMBER: HUMANS HAVE BEEN MODIFYING PLANTS AND ANIMALS--AND FUNGI--FOR AS LONG AS THEY'VE BEEN AGRICULTURISTS. CROSSBREEDING AND EVEN INDUCED MUTATIONS ARE RESPONSIBLE FOR CREATING ALL USEFUL DOMESTICATES FROM THEIR WILD ANCESTRY.
LET'S LOOK AT PLANT TRANSGENICS FIRST.
IT ALL STARTS WITH RECOMBINANT DNA TECHNOLOGY: THE GENE TO BE INSERTED--THE TRANSGENE--IS FOUND OR ALTERED, AND, USING AN APPROPRIATE VECTOR, INTRODUCED INTO A HOST PLANT'S DNA.
AGROBACTERIUM, WHICH ENTERS PLANT CELLS TO CAUSE TUMOR-LIKE GROWTHS CALLED GALLS, IS A COMMON, USEFUL VECTOR. LIKE VIRAL OR BACTERIAL VECTORS ALREADY DISCUSSED, AGRO HAS ITS UNDESIRABLE GENES CUT FROM ITS DNA PLASMID AND REPLACED WITH THE TRANSGENE.
AGROBACTERIUM
TRANSGENE DNA
ORIGINAL SOURCES
RECOMBINANT AGROBACTERIUM
REPLICATED AGROBACTERIA

HOST PLANT CELLS ARE INJECTED WITH THE ALTERED AGRO. A PERCENTAGE OF THE INFECTED CELLS GROW INTO MATURE PLANTS, WHICH CARRY THE TRANSGENE SUCCESSFULLY THROUGHOUT THE ENTIRE ORGANISM...
I FEEL TINGLY ALL OVER!
...AND EVEN PASS IT ON TO THEIR OFFSPRING.
AGRICULTURAL PRODUCERS ARE EXCITED BY THE POSSIBILITIES OF HERBICIDE- AND PESTICIDE-RESISTANT CROPS CREATED VIA TRANSGENICS. AN HERBICIDE-RESISTANT SWEET POTATO IS ONE RECENT SUCCESS.
JUST SO LONG AS IT ISN'T PEST-RESISTANT!
THE IDEA BEING THAT THESE GENETICALLY "IMPROVED" CROPS COULD REMAIN UNAFFECTED BY EVER MORE POTENT WEED AND PEST CONTROLS.
SO--COUGH-- WE CAN BE SMOTHERED WITH EVEN MORE DEADLY CHEMICALS?
KAFF!
EVEN MORE ASTONISHING-- AND PROBLEMATIC-- IS THE MEDICINAL POTENTIAL BEING EXPLORED.
HUMAN GENES PRODUCING TUMOR-FIGHTING ANTIBODY PROTEINS HAVE BEEN SUCCESSFULLY TRANSFERRED INTO A CROP OF CORN...
THE THINGS WE DOMESTICATES DO FOR YOU...
...AND TOMATOES IN A RUSSIAN LABORATORY. THEY NOW CARRY FRAGMENTS OF HIV DNA, PRODUCING A VACCINE THAT, WHEN EATEN, PROMPTS HUMAN GENES TO EXPRESS EFFECTIVE HIV ANTIBODIES!
AND WE MAKE A FINE SALSA.
UNFORTUNATELY, THESE "PLANTIBODIES" ARE ALWAYS SET TO "ON"--THEY NEVER STOP EXPRESSING.
THIS, COUPLED WITH THE ABILITY OF PLANTS TO TRANSFER THEIR GENES BETWEEN SPECIES, ALLOWS FOR THE UNCOMFORTABLE POSSIBILITY THAT ENGINEERED TRANSGENES MAY ESCAPE FROM DESIGNATED CROPS AND INFECT LARGER, UNCONTROLLED ENVIRONMENTS.
I FEEL TINGLY ALL OVER!

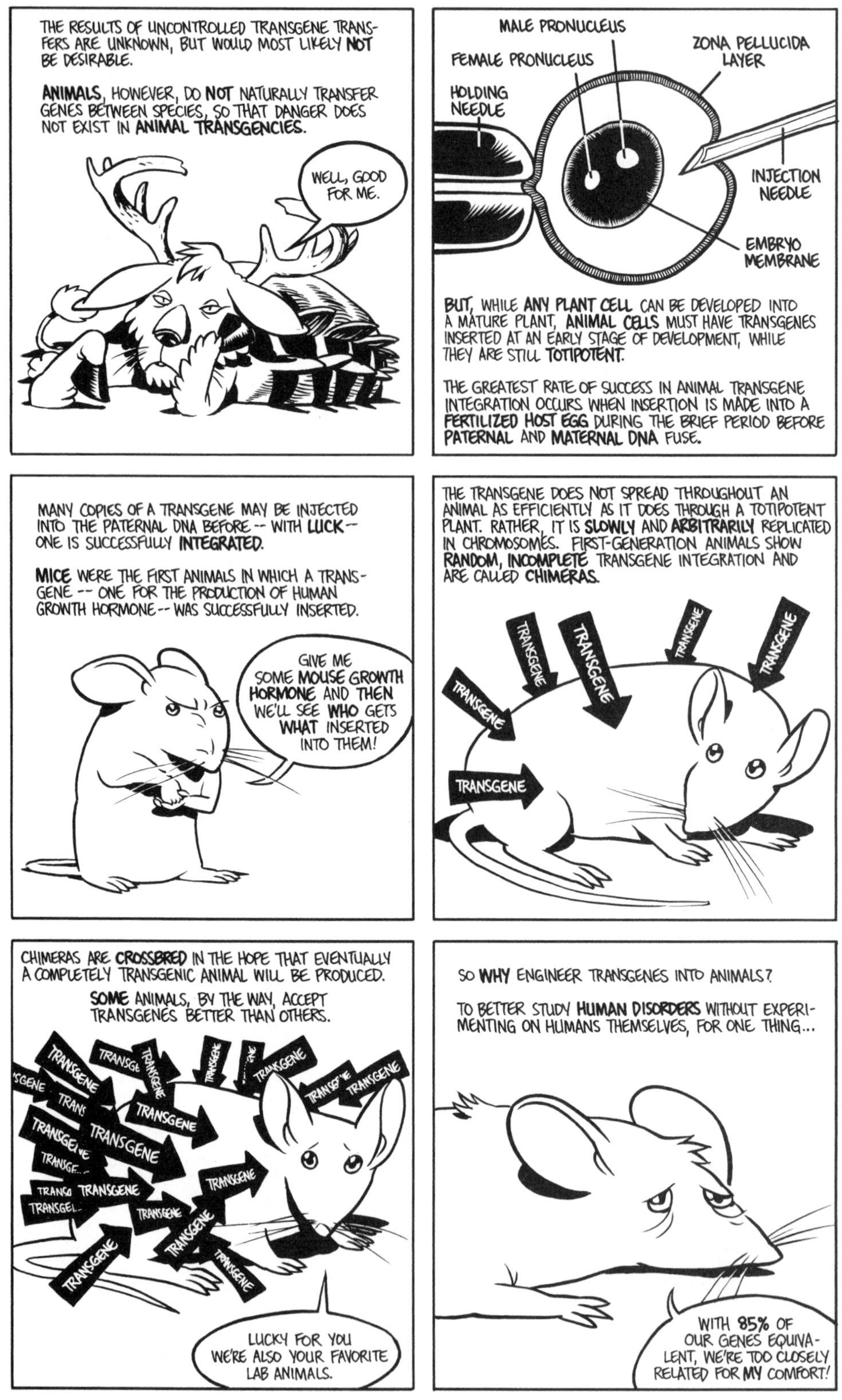
THE RESULTS OF UNCONTROLLED TRANSGENE TRANSFERS ARE UNKNOWN, BUT WOULD MOST LIKELY NOT BE DESIRABLE.
ANIMALS, HOWEVER, DO NOT NATURALLY TRANSFER GENES BETWEEN SPECIES, SO THAT DANGER DOES NOT EXIST IN ANIMAL TRANSGENCIES.
WELL, GOOD FOR ME.
MALE PRONUCLEUS
FEMALE PRONUCLEUS
ZONA PELLUCIDA LAYER
HOLDING NEEDLE
INJECTION NEEDLE
EMBRYO MEMBRANE
BUT, WHILE ANY PLANT CELL CAN BE DEVELOPED INTO A MATURE PLANT, ANIMAL CELLS MUST HAVE TRANSGENES INSERTED AT AN EARLY STAGE OF DEVELOPMENT, WHILE THEY ARE STILL TOTIPOTENT.
THE GREATEST RATE OF SUCCESS IN ANIMAL TRANSGENE INTEGRATION OCCURS WHEN INSERTION IS MADE INTO A FERTILIZED HOST EGG DURING THE BRIEF PERIOD BEFORE PATERNAL AND MATERNAL DNA FUSE.
MANY COPIES OF A TRANSGENE MAY BE INJECTED INTO THE PATERNAL DNA BEFORE -- WITH LUCK -- ONE IS SUCCESSFULLY INTEGRATED.
MICE WERE THE FIRST ANIMALS IN WHICH A TRANSGENE -- ONE FOR THE PRODUCTION OF HUMAN GROWTH HORMONE -- WAS SUCCESSFULLY INSERTED.
GIVE ME SOME MOUSE GROWTH HORMONE AND THEN WE'LL SEE WHO GETS WHAT INSERTED INTO THEM!
THE TRANSGENE DOES NOT SPREAD THROUGHOUT AN ANIMAL AS EFFICIENTLY AS IT DOES THROUGH A TOTIPOTENT PLANT. RATHER, IT IS SLOWLY AND ARBITRARILY REPLICATED IN CHROMOSOMES. FIRST-GENERATION ANIMALS SHOW RANDOM, INCOMPLETE TRANSGENE INTEGRATION AND ARE CALLED CHIMERAS.
TRANSGENE
TRANSGENE
TRANSGENE
TRANSGENE
TRANSGENE
TRANSGENE
CHIMERAS ARE CROSSBRED IN THE HOPE THAT EVENTUALLY A COMPLETELY TRANSGENIC ANIMAL WILL BE PRODUCED.
SOME ANIMALS, BY THE WAY, ACCEPT TRANSGENES BETTER THAN OTHERS.
TRANSGENE
LUCKY FOR YOU WE'RE ALSO YOUR FAVORITE LAB ANIMALS.
SO WHY ENGINEER TRANSGENES INTO ANIMALS?
TO BETTER STUDY HUMAN DISORDERS WITHOUT EXPERIMENTING ON HUMANS THEMSELVES, FOR ONE THING...
WITH 85% OF OUR GENES EQUIVALENT, WE'RE TOO CLOSELY RELATED FOR MY COMFORT!

...AND, SIMILAR TO PLANT TRANSGENICS, THE HOPE IS THAT HEALTHIER, CHEAPER, AND MORE CONSISTENT FOODS AND MEDICINES CAN BE PRODUCED.
A TRANSGENE CODING FOR A SPECIFIC ENZYME CONVERTER, INSERTED INTO PIGS, CONVERTS DIET-HARMFUL OMEGA-6 FATTY ACIDS INTO HEALTHY OMEGA-3 FATTY ACIDS!
HEY, WE'RE TALKING HEALTHY BACON HERE!
COWS NOW PRODUCE HUMAN PROTEIN C IN THEIR MILK. GOOD NEWS FOR HEMOPHILIACS AND FOR STROKE AND HEART ATTACK VICTIMS!
TRANSGENICS ALSO SHOWS PROMISE AS A MEANS OF CONTROLLING DISEASE-BEARING VECTORS.
TRANSGENES INSERTED INTO MALARIAL MOSQUITOES CAN RENDER THE INSECTS INCAPABLE OF CARRYING THE PARASITE RESPONSIBLE FOR THAT DREAD DISEASE.
GREAT! WHO WANTS PARASITES?
AND TRANSGENES FROM THREE DIFFERENT ANIMAL SOURCES HAVE BEEN INSERTED INTO YEAST TO HELP FABRICATE A DRUG TO FIGHT MALARIA.
THIS PARTICULAR PROJECT HAS ALREADY PROVEN A SUCCESS!
HMMM... MAYBE ANTI-MALARIAL RYE SOMEDAY...
ANIMAL TRANSGENES HAVE ALSO BEEN PUT TO SOME RATHER ODDBALL USES.
ZEBRAFISH, LACED WITH THE JELLYFISH GENE FOR FLUORESCENCE, FOR INSTANCE. THE INSERTED GENE IS USED AS A MARKER--AN ENVIRONMENTAL ALERT SYSTEM.
THE TRANSGENE IS MANIPULATED TO EXPRESS--TO FLUORESCE--ONLY WHEN POLLUTANTS IN ITS ZEBRAFISH HOST'S WATERS REACH A SPECIFIED UNHEALTHY LEVEL.
EVERYONE OUT OF THE POOL!

MANY HUMANS ARE UNDERSTANDABLY NERVOUS ABOUT THE POTENTIAL DANGERS GENETICALLY MODIFIED ORGANISMS MAY PRESENT.
BUT THAT NERVOUSNESS IS NOTHING COMPARED TO THE SWAMP OF FEAR AND ETHICAL CONTROVERSY SWIRLING AROUND ANOTHER ASPECT OF GENETIC ENGINEERING...
...CLONING TECHNOLOGY!
CLONES!!
A CLONE IS AN EXACT DUPLICATE OF AN ORGANISM AND, AS WITH THE POPULAR NOTIONS OF MUTANTS, CLONES HAVE BECOME A BYWORD FOR SCIENCE RUN AMOK...
...THIS DESPITE THE FACT THAT CLONES HAVE BEEN A BIG PART OF LIFE ON EARTH SINCE ITS BEGINNINGS.
YOUR OLD FRIENDS, BACTERIA, FOR INSTANCE, ARE A SIZABLE MAJORITY OF CLONES!
A GOOD DEAL OF VEGETATION REPRODUCES WITHOUT BENEFIT OF SEX, PRODUCING OFFSPRING THAT ARE GENETICALLY IDENTICAL TO THEIR SINGLE PARENT.
ASEXUAL REPRODUCTION IS CLONING!
SOME ANIMALS CAN DO IT THIS WAY, TOO. IN THIS KINGDOM, ASEXUAL REPRODUCTION IS COMMONLY REFERRED TO AS PARTHENOGENESIS.
'SNUTHIN' NEW!
LITTLE FIRE ANT
HYDRA
WITHIN HUMAN CULTURE, CLONING TECHNIQUES HAVE BEEN USED SINCE THE EARLIEST DAYS OF AGRICULTURE. FRUIT FARMERS PASS CONSISTENT, DESIRABLE TRAITS FROM ONE GENERATION OF CROP TO THE NEXT BY GRAFTING A CUTTING FROM THE PREVIOUS GENERATION ONTO NEW ROOT STOCK.
THE TRAITS OF THE OLD STOCK ARE TRANSFERRED TO THE NEW AS THE TWO FUSE AND THE DNA OF THE OLD INFORMS THE NEW.
GRAFTING IS CLONING!

JOHN CHAPMAN WAS A PIONEER OF THE AMERICAN WEST WHO GAVE THE NEW WORLD ITS OWN HEALTHY DIVERSITY OF APPLE VARIETIES. POPULARLY KNOWN AS JOHNNY APPLESEED, HE HAS BEEN MYTHOLOGIZED AND PRACTICALLY CANONIZED AS AN AMERICAN FOLK HERO. BUT THE ROUGH-HEWN FACTS OF WHAT HE ACCOMPLISHED ARE OFTEN WHITEWASHED TO THE POINT WHERE AN IMPORTANT LESSON IS LOST.
XXX
XXX
HOW THE WEST WAS SEEDED
(instead of cloned)
CHAPMAN HELPED DOMESTICATE THE AMERICAN FRONTIER IN THE EARLY YEARS OF THE 19th CENTURY BY PLANTING APPLE SEEDS WELL IN ADVANCE OF THE WESTWARD-SPREADING POPULACE. HE WAS A SORT OF LAND SPECULATOR -- HE WOULD SCOPE OUT NEW HORIZONS, DETERMINE WHERE SETTLEMENTS COULD BE MADE, AND THERE PLANT NEW ORCHARDS. BY THE TIME THE SETTLERS ARRIVED, AN ORCHARD WOULD BE WAITING. HE DID VERY WELL AT HIS BUSINESS (ALTHOUGH HIS ECCENTRIC, VAGABOND LIFESTYLE BELIED HIS FINANCIAL WORTH).
APPLE TREES ARE, BY NATURE, THE EXTREME OPPOSITE OF A TRUE-BREEDING PLANT. WHEN THEY ARE PLANTED FROM SEED, ONLY VERY RARELY WILL THE FRUIT OF THE OFFSPRING RESEMBLE THAT OF THE PARENT. APPLE TREES, NATIVES OF CENTRAL ASIA, ARE ECOLOGICALLY SUCCESSFUL PLANTS BECAUSE OF THIS STRATEGY OF EXTREME DIVERSITY.
HUMANS, HOWEVER, DESIRE A HIGH DEGREE OF CONSISTENCY IN THEIR CROPS. THE CHINESE DISCOVERED GRAFTING 4,000 YEARS AGO, AND IT WAS THIS CLONING TECHNIQUE THAT ALLOWED AGRICULTURISTS TO CREATE CONSISTENT VARIETIES OF APPLES. THE PRACTICE OF GRAFTING TO MAINTAIN STANDARDIZED FRUITS WAS PASSED ON TO EUROPEANS, WHO, IN TURN, BROUGHT BOTH THE APPLE AND THE PRACTICE TO THE WESTERN HEMISPHERE.
GRAFTED OLD-WORLD APPLE TREES DID NOT TEND TO FARE WELL UNDER AMERICAN ENVIRONMENTAL CONDITIONS. IT TOOK MEN LIKE CHAPMAN, WHO WERE WILLING TO GO BACK TO SEED, TO FIRMLY ESTABLISH THE EXOTIC APPLE AS A ROBUST, NATURALIZED TREE IN AMERICA. CHAPMAN WAS A DEEPLY RELIGIOUS MAN (ALTHOUGH THE EXACT NATURE OF HIS SPIRITUAL BELIEFS IS UP FOR DEBATE) WHO VIEWED GRAFTING AS UNNATURAL AND WOULD HAVE NOTHING TO DO WITH IT.
OF COURSE, THE APPLES FROM SEED THAT CHAPMAN PLANTED FOLLOWED THEIR TRUE NATURE AND EXPRESSED THEIR INCREDIBLE GENETIC DIVERSITY, PRODUCING A VAST MAJORITY OF BITTER OR SOUR FRUITS NOT PLEASANT TO HUMAN TASTE. SO, WHAT GOOD WERE ALL THOSE ORCHARDS THAT CHAPMAN PLANTED?
XXX
NOT MUCH -- IF YOU ARE CONSIDERING APPLES AS HUMANS DO TODAY. BUT IN CHAPMAN'S TIME APPLES WERE PRIMARILY USED TO MAKE AN ALCOHOLIC BEVERAGE -- CIDER. IT WAS MORE THAN A MERE INTOXICANT, TOO -- WITH WATER QUALITY BEING AT BEST AN IFFY PROPOSITION, CIDER WOULD BE A MUCH PREFERRED ALTERNATIVE TO CHOLERA.
APPLES WERE LARGELY REGARDED AS CIDER STOCK UNTIL THE 1920s, WHEN THE ANTI-ALCOHOL TEMPERANCE MOVEMENT THAT CULMINATED IN PROHIBITION DEMONIZED ALL DRINKS ALCOHOLIC. ONLY THEN DID THE APPLE-GROWING INDUSTRY REINVENT ITSELF AS PURVEYORS OF A TASTY, NUTRITIOUS FRUIT TO BE EATEN RAW, COINING THE PHRASE "AN APPLE A DAY KEEPS THE DOCTOR AWAY."
BUT NOW IT WAS TO THE INDUSTRY'S ADVANTAGE TO PRODUCE CONSISTENT FRUIT THAT TASTED RELIABLY SWEET AND COULD BE EFFICIENTLY SHIPPED. GRAFTING BECAME STANDARD PROCEDURE FOR PROPAGATING AMERICAN APPLES, AND DIVERSITY WAS REDUCED TO THE HANDFUL OF CLONES SEEN IN SUPERMARKETS TODAY, ALL RELYING ON EVER-INCREASING CHEMICAL ASSISTANCE TO BATTLE CATASTROPHIC DEVASTATION FROM THE RAPIDLY EVOLVING INSECTS, FUNGI, AND MICROBES THAT HAVE GAINED THE EVOLUTIONARY UPPER HAND.
IF THERE IS A GREAT APPLE ORCHARD IN THE SKY, JOHN CHAPMAN MUST BE LOOKING DOWN, NOT AT ALL HAPPY.

A COMMERCIAL VARIETY OF APPLE -- A MACINTOSH, FOR INSTANCE -- THAT A HUMAN EATS TODAY IS GENETICALLY IDENTICAL TO A MACINTOSH EATEN 30 YEARS AGO!
STILL TASTES FRESH TO ME.
BUT THAT'S MERE RECENT HISTORY! THERE ARE GRAPE CULTIVARS IN EUROPE THAT HAVE BEEN PROPAGATED BY CLONING FOR OVER 2,000 YEARS!
AND YOU BET WE'RE FEELING OUR AGE...
THE HUMAN SPECIES ITSELF IS RIFE WITH CLONES -- BUT HUMANS REFER TO THEM AS IDENTICAL TWINS.
WE HAVE ALWAYS LIVED AMONG YOU, BIDING OUR TIME...
WHOA! Wait a minute -- this clone business...
...This means we Squinch are CLONES, right?
WE reproduce without benefit of sex...
...thank goodness...
THAT'S RIGHT -- IT'S PARTHENOGENESIS FOR US.
WE ARE CLONES.
AND A VERY HYGIENICALLY SOUND STRATEGY, IN MY OPINION.
So what could humans POSSIBLY find frightening about CLONES?!
I mean, LOOK at us! We're ADORABLE!

?
WELL, LET ME BE PERFECTLY CLEAR-- IT WAS, AS USUAL, ADVANCES IN SCIENTIFIC KNOWLEDGE AND TECHNOLOGY THAT CREATED THE UPROAR.
IN 1952, HUMANS FIRST ARTIFICIALLY CLONED AN ANIMAL-- A FROG TADPOLE--FROM EMBRYONIC CELLS.
BUT IT WASN'T UNTIL 1996 AND THE FIRST SUCCESSFUL CLONING OF A MUCH MORE COMPLEX ANIMAL -- A MAMMAL-- THAT THE FUROR BEGAN IN EARNEST.
MAMMALS ARE SO CUTE! THEY'RE JUST LIKE US!
DOLLY THE SHEEP WAS THE FIRST MAMMAL CLONED FROM AN ADULT CELL THAT LIVED PAST BIRTH.
HER COST WAS EXTRAORDINARILY HIGH. SHE WAS THE ONLY SUCCESS IN A PROJECT ATTEMPTING TO PRODUCE CLONES IN 277 SHEEP EGGS.
DOLLY DIED, PREMATURELY, AT AGE SIX-- A DISAPPOINTINGLY BRIEF LIFE FOR A SHEEP. WHETHER THE CAUSE OF DEATH WAS RELATED TO HER CLONED NATURE OR NOT IS STILL A MATTER OF DEBATE.
SINCE DOLLY, THERE HAVE BEEN OTHER SUCCESSES-- AND FAILURES. BUT MORE FAILURES THAN SUCCESSES.
THE EXCITEMENT OVER CLONING HAS, IN MANY REGARDS, BEEN SUPPLANTED BY ADVANCES IN RECOMBINANT DNA TECHNOLOGY.
TRUTH BE TOLD, RECOMBINANT DNA TECHNOLOGY IS REALLY JUST CLONING ON A MOLECULAR LEVEL!
REPRODUCTIVE CLONING -- REPRODUCING WHOLE ANIMALS-- HAS REMAINED TOO EXPENSIVE AND PROBLEMATIC TO MAKE GOOD MEDICAL OR COMMERCIAL SENSE...
... ALTHOUGH THERE ARE PARTIES WHO REMAIN HOPEFUL.
WHAT CARNIVORE WOULDN'T WANT A GUARANTEED PERFECT GRADE A SIRLOIN STEAK FOR EVERY MEAL?
MAYBE EVEN WITH OMEGA-3 FATTY ACIDS!

ANOTHER USE FOR CLONING THAT MAY HAVE A FUTURE IS THAT OF THERAPEUTIC TISSUE REPLICATION-- THE CLONING OF INDIVIDUAL ORGANS FOR TRANSPLANT PURPOSES.
IT BEATS WAITING FOR A DONOR.
HUMAN EGG CELL
THE SCENARIO WORKS LIKE THIS: DNA FROM AN INDIVIDUAL NEEDING A REPLACEMENT ORGAN IS INSERTED INTO A HUMAN EGG STRIPPED OF ITS NUCLEUS.
TISSUE CLONED FROM THE PATIENT'S DNA WOULD AVOID THE DANGER OF DONOR-ORGAN REJECTION BY THE IMMUNE SYSTEM.
AFTER THE EGG, WITH THE TRANSFERRED DNA, BEGINS TO DIVIDE, EMBRYONIC STEM CELLS -- CELLS THAT ARE AT THIS STAGE CAPABLE OF DEVELOPING INTO ANY TYPE OF TISSUE THAT THE BODY NEEDS--ARE HARVESTED AND GUIDED TO GENERATE THE SPECIFIC ORGAN TISSUE NEEDED.
SUDDENLY, I FEEL LIKE BECOMING A HEART...
THE NEW ORGAN WOULD BE A CLONE OF THE PATIENT'S ORIGINAL.
YEP--THE SPITTING IMAGE OF MY ORIGINAL EQUIPMENT!
IF THIS TECHNOLOGY BECOMES FEASIBLE, THE SAFETY AND SUCCESS RATE FOR TRANSPLANTATION WOULD BE GREATLY INCREASED, WHILE THE NEED FOR ORGAN DONATION WOULD BE DRASTICALLY REDUCED.
BUT...
...AS RESEARCH IS NOW CONDUCTED THROUGH A PROCESS THAT KILLS THE HUMAN EGG--A POTENTIAL HUMAN LIFE-- WHEN STEM CELLS ARE HARVESTED, VARIOUS INTEREST GROUPS HAVE SOUGHT TO RESTRICT OR OUTLAW THEIR USE.
THE ETHICAL OR SUPERSTITIOUS OBJECTIONS THESE INTEREST GROUPS RAISE PLACE LIMITATIONS ON THE CONTINUATION OF STEM CELL RESEARCH...
...ALTHOUGH THE OBJECTIONS MIGHT BE OVERCOME BY RECENT DEVELOPMENTS INDICATING THAT THERE MAY BE A MEANS OF HARVESTING THESE VALUABLE CELLS WITHOUT DAMAGING THE EGG.
ONLY TIME WILL TELL HOW THIS DRAMA OF MEDICAL ADVANCEMENT VERSUS MORAL CONCERN ENDS!

WHILE TISSUE CLONING STILL SHOWS POTENTIAL, HOWEVER, SCIENTIFIC INTEREST IN THE POSSIBILITIES OF REPRODUCTIVE CLONING SEEMS TO HAVE WANED. ALTHOUGH NEW SPECIES ARE INDEED CLONED EVERY YEAR...
...THE ENORMOUS EXPENSE AND THE HIGH RATE OF FAILURE CONTINUE TO BE ROADBLOCKS.
A QUALIFIED BRIGHT SPOT: CONSERVATIONISTS HAVE LOOKED TO CLONING AS A LAST-DITCH, WORST-CASE SCENARIO FOR SAVING SPECIES ON THE BRINK OF EXTINCTION.
WELL, IT WORKED FOR ME...
YEAH, BUT TO LIVE ON WITHOUT THE ENVIRONMENT IN WHICH I EVOLVED...
...I DON'T KNOW...
MOUFLON
SUMATRAN TIGER
AND FINALLY: THE IDEA THAT GENETIC ENGINEERING WILL PRODUCE THE PERFECT BABY-TO-ORDER?
THE NOTION THAT CLONING WILL KEEP A BELOVED PET WITH US FOREVER?
THAT'S NOT GOING TO HAPPEN. EVER.
CLONES ARE EXACT DUPLICATES IN GENOTYPE ONLY. THERE ARE FAR TOO MANY ENVIRONMENTAL AND MUTATIONAL FACTORS AT WORK IN PRODUCING THE PHENOTYPE OF A COMPLEX ORGANISM TO EVER PRODUCE AN EXPRESSLY CRAFTED DUPLICATE-ON-DEMAND.
WATERGATE
The Fall of Richard M Nix
I CAN'T BELIEVE HOW MY ORIGINAL BEHAVED! WHAT KIND OF UPBRINGING DID HE HAVE?
TELL me about it! None of MY offspring are exactly chips off the old block.
BUT WE ARE ALL GRATEFUL FOR THAT, I ASSURE YOU.
All joking aside...
...I am well pleased with what you have taught me today, Bloort 183. Your mission to Earth was undeniably worthwhile. You have provided us with some important lessons--
--as alien and hard to digest as they may be.
But, tell me, just how is it that we Squinch have never explored our own biology so thoroughly?

MAYBE, O WAKENING GIANT, BECAUSE WE EVOLVED SO **SLOWLY** WHILE SOMEHOW MANAGING TO AVOID A MASSIVE CRISIS SUCH AS NOW CONFRONTS OUR SPECIES.
BECAUSE WE HAVE BEEN **SO** STABLE, MAYBE THE NOTION OF CHANGE NEVER OCCURRED TO US.
BUT EVEN THOUGH WE ARE ASEXUAL BEINGS WITHOUT THE MECHANICS FOR NIMBLE ADAPTATION, WE **HAVE** EVOLVED...
...SLOWLY AND PONDEROUSLY, THROUGH GRADUAL, PERSISTENT MUTATION.
WE **DO** GENETICALLY CHANGE.
WHAT WE MUST DO **NOW** IS FIND A WAY OF **HARNESSING** WHAT WE HAVE **WITHIN** OUR BIOLOGY IN ORDER TO BEAT THIS CRISIS!
Well, the **HUMANS** are developing a **VARIETY** of strategies for curing their genetic ills.
BUT THOSE WON'T NECESSARILY WORK FOR **US**...
WE MUST FIRST EXPLORE **OUR** GENETIC HISTORY -- THE GENETIC HISTORY OF **OUR** WORLD -- TO UNDERSTAND WHERE **WE** COME FROM.
TO UNDERSTAND WHY WE ARE NOW FAILING BEFORE THIS...UNFORTUNATE **DISORDER**.
Very well, then. We must learn our genetic past.
Even down to the -- *sigh* -- Frippurvian flogworm.
So tell us, Bloort -- just how **ARE** the humans uncovering **THEIR** history?

CHAPTER 5
Applying All That Stuff-- Into the Future Through the Past
MODERN HUMAN BEINGS-- HOMO SAPIENS IN SCIENTIFIC NOMENCLATURE-- ARE A SINGLE SPECIES, BUT ONE WITH PLENTY OF VARIATION.
ALL THIS VARIATION IS LARGELY PHENOTYPICAL, HOWEVER. EVERY HUMAN'S PERSONAL GENOME-- WHICH CONTAINS THEIR GENOTYPE-- DIFFERS FROM ANOTHER'S BY ONLY A FEW HUNDREDTHS OF A PERCENTAGE POINT.
REMEMBER-- 98% OF THE GENES OF HUMANS AND THEIR CLOSEST LIVING RELATIVES, CHIMPANZEES, ARE PRACTICALLY IDENTICAL!
WHILE HUMANS BECAME ANATOMICALLY MODERN SOMEWHERE AROUND 150,000 TO 200,000 YEARS AGO, WITH THEIR MODERN BEHAVIOR PATTERNS APPEARING ONLY ABOUT 50,000 YEARS AGO, THE VARIATIONS WE SEE IN DIFFERENT POPULATIONS SCATTERED ACROSS THE EARTH HAVE APPEARED MUCH MORE RECENTLY.
AND DIFFERING PHYSICAL CHARACTERISTICS ONCE CONSIDERED TO DEFINE RACES (AN OUTMODED CONCEPT) APPEARED AS RECENTLY AS ONLY 5,000 YEARS AGO!
BUT THE VARIATIONS IN THE HUMAN SPECIES ARE SIGNIFICANT. THEY REFLECT THE FACT THAT HUMANS MIGRATED FROM A SINGLE POINT OF ORIGIN TO ESTABLISH GENETICALLY DIFFERENTIATED POPULATIONS AROUND THE GLOBE.
EACH ISOLATED POPULATION FACED ITS OWN ENVIRONMENTAL AND CULTURAL CHALLENGES AND IN RESPONSE EVOLVED ITS OWN SET OF UNIQUE GENETIC ASSETS -- AND LIABILITIES.
DNA REVEALS THE STILL-DEVELOPING STORY OF THE HUMAN SPECIES!

HUMANS HAVE A PRETTY GOOD IDEA OF THEIR HISTORY GOING BACK 5,000 YEARS, WHICH INCLUDES THE EARLIEST KNOWN WRITTEN RECORDS.
1
6
3
THE MOST ANCIENT EVIDENCE THEY'VE YET DISCOVERED FOR HUMAN SETTLEMENTS IS ABOUT 40,000 YEARS OLD, BUT THE SPECIFICS GOT INCREASINGLY FOGGY ANY EARLIER THAN THAT...
...UNTIL GENETIC RESEARCHERS BECAME ANTHROPOLOGICAL DETECTIVES AND BEGAN PIECING TOGETHER THE PUZZLE OF EARLY HUMAN BIOLOGICAL, SOCIAL, AND CULTURAL EVOLUTION.
?
THE SEARCH FOR THE ORIGINS OF THE MODERN HUMAN SPECIES HAS BENEFITED GREATLY FROM TWO GENETIC FACTORS...
...ONE IS THE Y CHROMOSOME -- CARRIED BY MALES ONLY...
...AND THE OTHER IS THE DNA FOUND WITHIN MITOCHONDRIA, UNIQUE TO THAT CELLULAR ORGANELLE...
...AND PASSED EXCLUSIVELY FROM MOTHER TO CHILD.
TRACING HUMAN HISTORY THROUGH THE Y CHROMOSOME IS MADE EASIER BY THE FACT THAT ALL MEN TODAY CARRY A Y CHROMOSOME INHERITED FROM A SINGLE ANCESTOR!
WHY? BECAUSE THAT ANCIENT MAN'S LINEAGE OUTLASTED ALL OF HIS APPROXIMATELY 2,500 MALE COMPETITORS!
THROG'S MALE SEED IS GOING NOWHERE -- HE'S GIVEN LIFE ONLY TO DAUGHTERS!
BUT, AS WE'VE SEEN ELSEWHERE, INHERITED CHROMOSOMES ARE NOT IDENTICAL COPIES. MUTATIONS ARE CONSTANTLY OCCURRING WITHIN GENES.
ONCE A MUTATION APPEARS ON A Y CHROMOSOME, IT IS HANDED DOWN THROUGH SUCCESSIVE GENERATIONS, FATHER TO SON.
THIS Y CHROMOSOME IS YOUR INHERITANCE, SON. I RECEIVED IT FROM MY FATHER, BUT THERE HAVE BEEN SOME IMPROVEMENTS MADE...

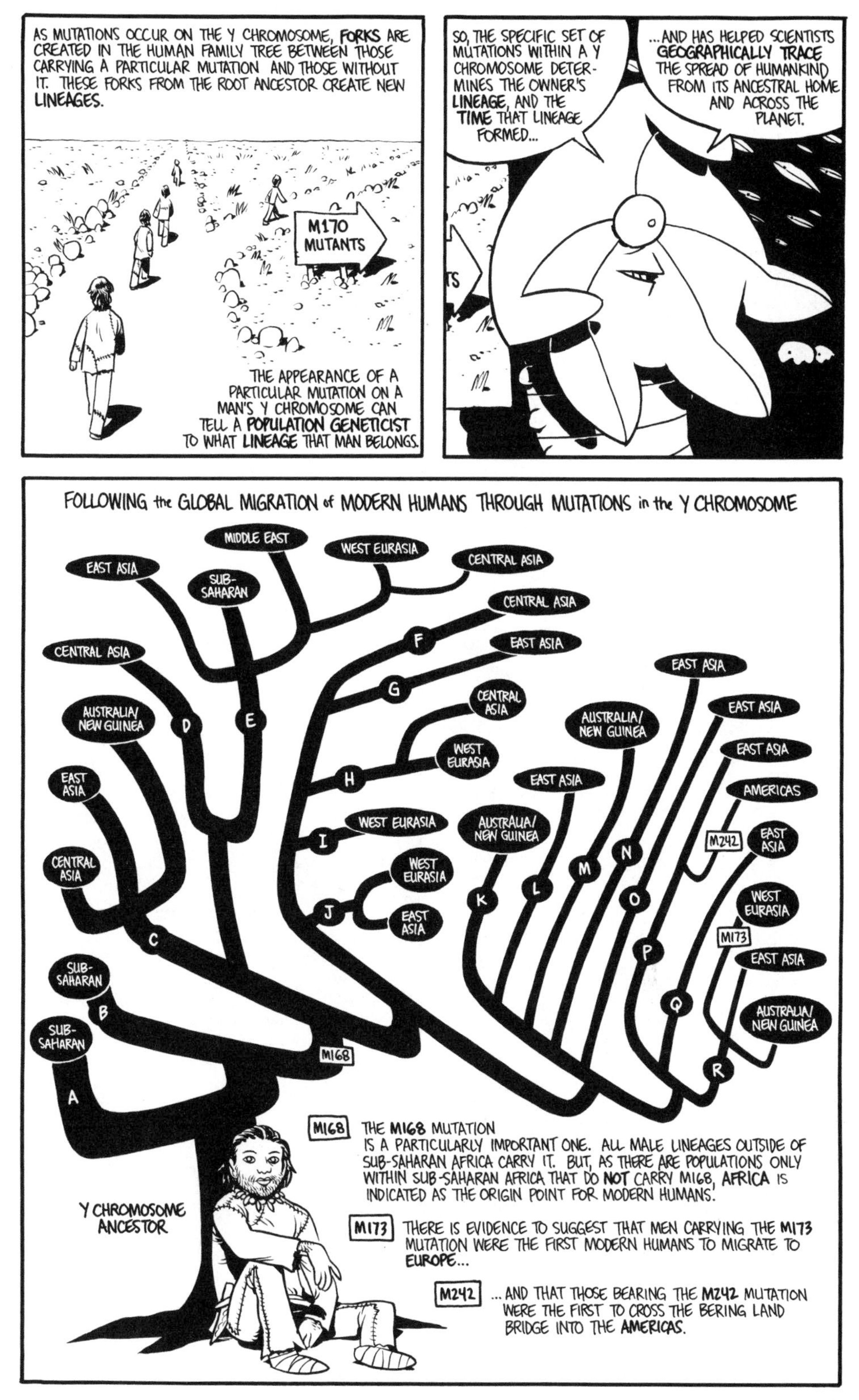
AS MUTATIONS OCCUR ON THE Y CHROMOSOME, FORKS ARE CREATED IN THE HUMAN FAMILY TREE BETWEEN THOSE CARRYING A PARTICULAR MUTATION AND THOSE WITHOUT IT. THESE FORKS FROM THE ROOT ANCESTOR CREATE NEW LINEAGES.
M170 MUTANTS
THE APPEARANCE OF A PARTICULAR MUTATION ON A MAN'S Y CHROMOSOME CAN TELL A POPULATION GENETICIST TO WHAT LINEAGE THAT MAN BELONGS.
SO, THE SPECIFIC SET OF MUTATIONS WITHIN A Y CHROMOSOME DETERMINES THE OWNER'S LINEAGE, AND THE TIME THAT LINEAGE FORMED...
...AND HAS HELPED SCIENTISTS GEOGRAPHICALLY TRACE THE SPREAD OF HUMANKIND FROM ITS ANCESTRAL HOME AND ACROSS THE PLANET.
FOLLOWING the GLOBAL MIGRATION of MODERN HUMANS THROUGH MUTATIONS in the Y CHROMOSOME
A
SUB-SAHARAN
B
SUB-SAHARAN
C
CENTRAL ASIA
EAST ASIA
AUSTRALIA/ NEW GUINEA
D
CENTRAL ASIA
EAST ASIA
E
SUB-SAHARAN
MIDDLE EAST
WEST EURASIA
CENTRAL ASIA
F
CENTRAL ASIA
G
EAST ASIA
H
CENTRAL ASIA
WEST EURASIA
I
WEST EURASIA
J
WEST EURASIA
EAST ASIA
K
AUSTRALIA/ NEW GUINEA
L
EAST ASIA
M
AUSTRALIA/ NEW GUINEA
N
EAST ASIA
O
EAST ASIA
P
EAST ASIA
M242
AMERICAS
Q
EAST ASIA
M173
WEST EURASIA
R
EAST ASIA
AUSTRALIA/ NEW GUINEA
M168
Y CHROMOSOME ANCESTOR
M168 THE M168 MUTATION IS A PARTICULARLY IMPORTANT ONE. ALL MALE LINEAGES OUTSIDE OF SUB-SAHARAN AFRICA CARRY IT. BUT, AS THERE ARE POPULATIONS ONLY WITHIN SUB-SAHARAN AFRICA THAT DO NOT CARRY M168, AFRICA IS INDICATED AS THE ORIGIN POINT FOR MODERN HUMANS.
M173 THERE IS EVIDENCE TO SUGGEST THAT MEN CARRYING THE M173 MUTATION WERE THE FIRST MODERN HUMANS TO MIGRATE TO EUROPE...
M242 ...AND THAT THOSE BEARING THE M242 MUTATION WERE THE FIRST TO CROSS THE BERING LAND BRIDGE INTO THE AMERICAS.

SCIENTISTS ARE ABLE TO CALCULATE THE RATE AT WHICH MUTATIONS ACCUMULATE. THE M168 MUTATION'S APPEARANCE-- SOON BEFORE MODERN HUMANS BEGAN THEIR SUCCESSFUL TREK OUT OF AFRICA--
M168
--CAN BE DATED TO ABOUT 44,000 YEARS AGO...
...WHILE THE GENETIC "ADAM" WHO CONTRIBUTED THE MODERN HUMAN Y CHROMOSOME IS ESTIMATED TO HAVE LIVED ABOUT 59,000 YEARS AGO.
THESE DNA-DERIVED DATES ARE ONLY APPROXIMATE, BUT THEY FIT NICELY WITH PALEONTOLOGICAL EVIDENCE THAT ALSO PLACES GREAT ADVANCES IN MODERN HUMAN BEHAVIOR AT AROUND 50,000 YEARS AGO.
WHAT'S A FEW THOUSAND YEARS? I'M STILL YOUR FATHER!
MITOCHONDRIAL DNA SEEMS TO CONFIRM THIS HISTORY OF HUMAN DISPERSION AS WELL.
LIKE THE Y CHROMOSOME LINEAGE, ALL MITOCHONDRIAL DNA COMES FROM A SINGLE ROOT SOURCE...
... A MITOCHONDRIAL "EVE" WHOSE LINEAGE OUTLASTED ALL HER COMPETING MATRIARCHAL LINEAGES.
LIKE CHROMOSOMAL DNA, MITOCHONDRIAL DNA COLLECTS MUTATIONS, ALLOWING SCIENTISTS TO CALCULATE DATES...
...ALTHOUGH WITH SOMEWHAT LESS ACCURACY THAN WITH THE Y CHROMOSOME, AS MITOCHONDRIAL DNA MUTATES MORE QUICKLY.
EVEN SO, THE TIME FRAME FOR THE MOTHER OF MODERN HUMAN MITOCHONDRIAL DNA CAN BE APPROXIMATED AT ABOUT 150,000 YEARS AGO, SIGNIFICANTLY OLDER THAN THE Y CHROMOSOME FATHER.
I AM THE MOTHER OF ALL MOTHERS! AND ALL FATHERS!
BASED ON THE NUMBER OF ALLELIC VARIATIONS SEEN IN BOTH CONTEMPORARY Y CHROMOSOMES AND MITOCHONDRIAL DNA AND THE KNOWN RATE OF MUTATION ACCUMULATION, THE SIZE OF THE ANCESTRAL MODERN HUMAN POPULATION IS ESTIMATED TO BE AROUND ONLY 5,000 INDIVIDUALS!
2,500 MALES...
...AND 2,500 FEMALES. AND ONLY ONE MALE Y CHROMOSOME LINEAGE AND ONE FEMALE MITOCHONDRIAL LINEAGE WILL SURVIVE TO YOUR TIME!

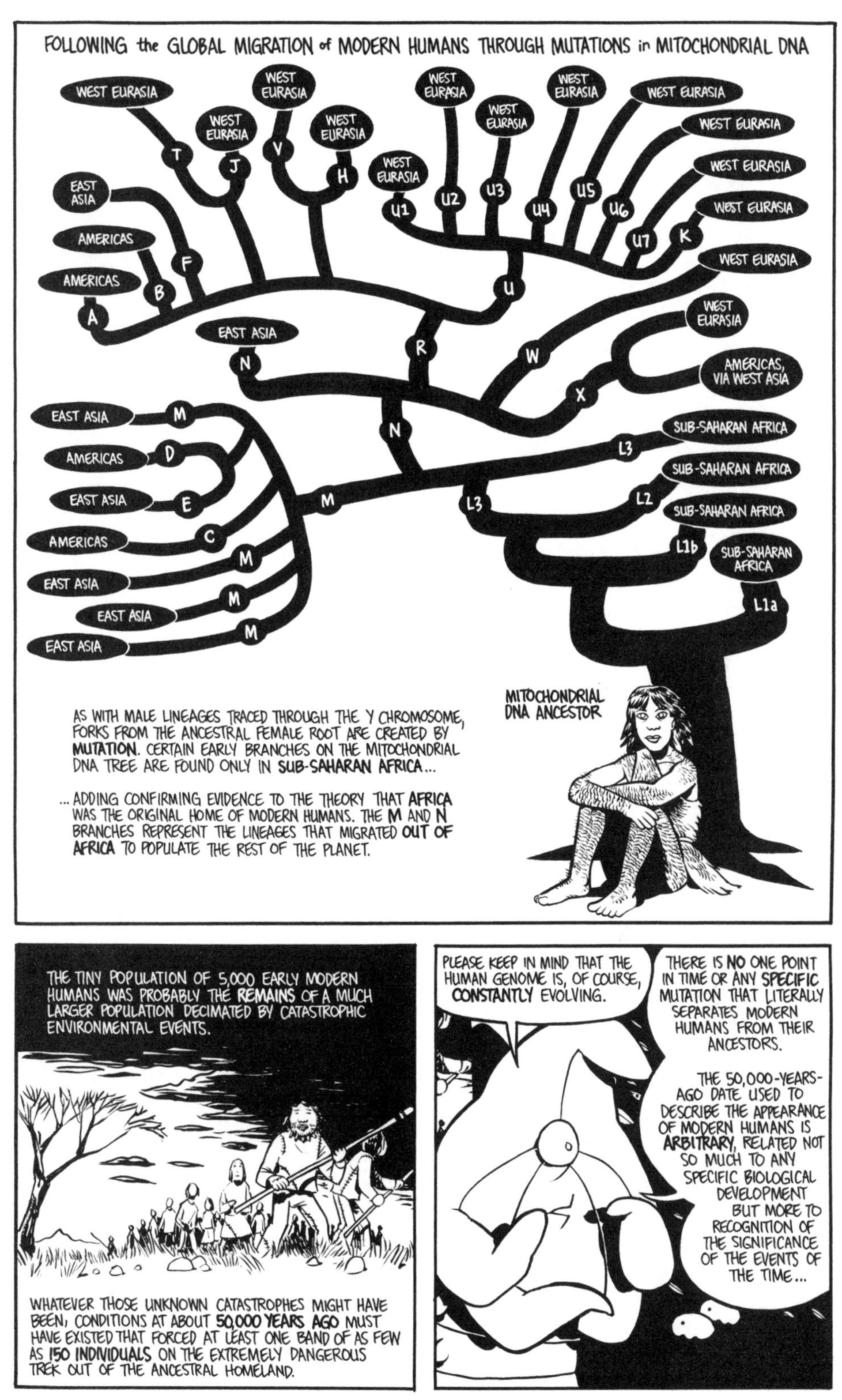
FOLLOWING the GLOBAL MIGRATION of MODERN HUMANS THROUGH MUTATIONS in MITOCHONDRIAL DNA
WEST EURASIA
T
WEST EURASIA
J
WEST EURASIA
V
WEST EURASIA
H
WEST EURASIA
U1
WEST EURASIA
U2
WEST EURASIA
U3
WEST EURASIA
U4
WEST EURASIA
U5
WEST EURASIA
U6
WEST EURASIA
U7
WEST EURASIA
K
EAST ASIA
F
AMERICAS
B
AMERICAS
A
U
WEST EURASIA
WEST EURASIA
R
EAST ASIA
N
W
AMERICAS, VIA WEST ASIA
X
EAST ASIA
M
AMERICAS
D
EAST ASIA
E
AMERICAS
C
M
EAST ASIA
M
EAST ASIA
M
EAST ASIA
M
N
SUB-SAHARAN AFRICA
L3
SUB-SAHARAN AFRICA
L2
L3
SUB-SAHARAN AFRICA
L1b
SUB-SAHARAN AFRICA
L1a
MITOCHONDRIAL DNA ANCESTOR
AS WITH MALE LINEAGES TRACED THROUGH THE Y CHROMOSOME, FORKS FROM THE ANCESTRAL FEMALE ROOT ARE CREATED BY MUTATION. CERTAIN EARLY BRANCHES ON THE MITOCHONDRIAL DNA TREE ARE FOUND ONLY IN SUB-SAHARAN AFRICA...
...ADDING CONFIRMING EVIDENCE TO THE THEORY THAT AFRICA WAS THE ORIGINAL HOME OF MODERN HUMANS. THE M AND N BRANCHES REPRESENT THE LINEAGES THAT MIGRATED OUT OF AFRICA TO POPULATE THE REST OF THE PLANET.
THE TINY POPULATION OF 5,000 EARLY MODERN HUMANS WAS PROBABLY THE REMAINS OF A MUCH LARGER POPULATION DECIMATED BY CATASTROPHIC ENVIRONMENTAL EVENTS.
WHATEVER THOSE UNKNOWN CATASTROPHES MIGHT HAVE BEEN, CONDITIONS AT ABOUT 50,000 YEARS AGO MUST HAVE EXISTED THAT FORCED AT LEAST ONE BAND OF AS FEW AS 150 INDIVIDUALS ON THE EXTREMELY DANGEROUS TREK OUT OF THE ANCESTRAL HOMELAND.
PLEASE KEEP IN MIND THAT THE HUMAN GENOME IS, OF COURSE, CONSTANTLY EVOLVING.
THERE IS NO ONE POINT IN TIME OR ANY SPECIFIC MUTATION THAT LITERALLY SEPARATES MODERN HUMANS FROM THEIR ANCESTORS.
THE 50,000-YEARS-AGO DATE USED TO DESCRIBE THE APPEARANCE OF MODERN HUMANS IS ARBITRARY, RELATED NOT SO MUCH TO ANY SPECIFIC BIOLOGICAL DEVELOPMENT BUT MORE TO RECOGNITION OF THE SIGNIFICANCE OF THE EVENTS OF THE TIME...

...MOST IMPORTANT, THE SUCCESSFUL TREK OUT OF AFRICA.
THERE IS ARCHAEOLOGICAL EVIDENCE OF EARLIER ATTEMPTS THAT FAILED, POSSIBLY THE RESULT OF DIRECT CONFLICT WITH HOSTILE FORCES OCCUPYING LANDS THAT BLOCKED THE ROUTE FROM THE HOMELAND.
NO COME BACK!
MOST LIKELY, THE HOSTILES WERE MODERN HUMANS' EXTINCT RELATIVE, *HOMO NEANDERTHALENSIS*-- NEANDERTHAL MAN.
WHY DID MODERN HUMANS FINALLY SUCCEED IN BREAKING OUT OF AFRICA?
GENETIC RESEARCH HAS DISCOVERED CHANGES IN THE HUMAN BRAIN THAT MOST LIKELY LED TO THE DEVELOPMENT OF IMPORTANT ADVANTAGES AT ABOUT THIS TIME: SOPHISTICATED LINGUISTICS AND SOCIAL STRUCTURES.
...AND SO, MY DEAR FELLOWS, I SAY WE MUST ALL HANG TOGETHER -- OR SURELY WE WILL ALL HANG SEPARATELY...
TO FIND THE ORIGINS OF THESE GENETIC ADVANCES WE NEED TO LOOK FURTHER BACK, TO THE TIME WHEN THE LINEAGES OF PROTO-HUMANS AND THEIR CLOSEST RELATIVES, CHIMPANZEES...
CHIMPANZEE
...SEPARATED FROM THEIR ANCIENT COMMON ANCESTOR.
ANCESTRAL PRIMATE
PROTO-HUMAN
HUMANS AND CHIMPS HAVE BETTER THAN 98% OF THEIR GENES ANALOGOUS OR HOMOLOGOUS. BY COMPARING THE NUMBER OF DIFFERENCES IN CORRESPONDING STRETCHES OF HUMAN AND CHIMP DNA, GENETICISTS ARE ABLE TO DATE THE SPLIT BETWEEN THE TWO SPECIES AT APPROXIMATELY 5 MILLION YEARS AGO.
GOODBYE! LET'S NOT LET OUR DIFFERENCES MAKE US STRANGERS!
SEVERE CLIMATIC SHIFTS AT THAT TIME WERE CAUSING MASSIVE ENVIRONMENTAL CHANGES IN AFRICA. INDIVIDUALS WITH MUTATIONAL VARIATIONS BEST SUITED TO COPING WITH THE NEW, MORE ARID CONDITIONS WERE MOST LIKELY TO SURVIVE-- AND SO PASS ON THEIR GENES TO FUTURE GENERATIONS.
OUT OF THE FOREST AND INTO THE GRASS! AN EXCELLENT EXAMPLE OF PRACTICAL EVOLUTION!
ONE OF THE MOST IMPORTANT MUTATIONS -- A REAL ASSET AS HUMANS BRANCHED OUT TO CONQUER A NEW ENVIRONMENTAL NICHE -- INVOLVED THE DEVELOPMENT OF THE BRAIN.
GENETICISTS, AGAIN COMPARING CHIMP AND HUMAN GENOMES, HAVE FOUND SPECIFIC STRETCHES OF DNA ENCOURAGING RAPID GENE MUTATION ONLY IN THE HUMAN!
I CAN'T BELIEVE WE USED TO HANG OUT WITH THOSE GUYS.

THESE HUMAN-ACCELERATED REGIONS,
OR HARS...
...OFFER CLUES TO HUMAN BRAIN EVOLUTION. MANY OF THE HARS ARE LOCATED NEAR GENES SPECIFICALLY INVOLVED WITH BRAIN DEVELOPMENT.
HAR 1 LIES WITHIN A GENE THAT CODES FOR AN RNA MOLECULE THAT MAY SUPERVISE BRAIN CONSTRUCTION IN THE WOMB.
PRIORITIZE ACCELERATED DEVELOPMENT
HAR 1-- SIGNIFICANTLY NOT PRESENT IN THE CHIMP GENOME-- MAY REPRESENT AN IMPORTANT MUTATION AT LEAST PARTIALLY RESPONSIBLE FOR HUMANKIND'S UNIQUE ADVANCEMENT INTO INTELLECTUAL SOPHISTICATION.
IMPROVED INTELLECTUAL CAPACITY QUITE POSSIBLY LED TO MUCH MORE COMPLEX SOCIAL STRUCTURE, INCLUDING MALE-FEMALE PAIR BONDING, AND LONG-TERM FAMILY RELATIONSHIPS THAT ENGENDERED A GREATER INTEREST IN GROUP WELFARE...
...AND AN EXTREMELY COHESIVE, COOPERATIVE SOCIAL SYSTEM, CAPABLE OF QUICKLY DISSEMINATING NEW CULTURAL DEVELOPMENTS. THIS WOULD HAVE BEEN AN IMPORTANT ADVANTAGE WHEN IT CAME TO RAISING FAMILIES, HUNTING, AND GATHERING FOOD...
...AND SURVIVING CONFLICTS WITH OTHER SPECIES!
?
TO UNDERSTAND THE IMPACT OF THESE ADVANCES IN HUMAN DEVELOPMENT, WE NEED TO JUMP FORWARD TO A POINT IN TIME...
...WHERE THE PROTO-HUMAN TREE FORKED. THE PROTO-HUMAN LINEAGE DID NOT LEAD SOLELY TO MODERN HUMANS--
--HOMO SAPIENS-- BUT BRANCHED INTO OTHER SPECIES, INCLUDING HOMO ERECTUS AND HOMO NEANDERTHALENSIS.
BOTH OF THESE HUMAN SPECIES HAD MANAGED TO MIGRATE OUT OF AFRICA AS EARLY AS 1.8 MILLION YEARS AGO, POPULATING EUROPE AND ASIA AS THEY FOLLOWED THEIR OWN EVOLUTIONARY PATHS.
ME GO NORTH.
ME WALK EAST.

IT WAS POSSIBLY THE HEAVILY MUSCLED, INTELLIGENT NEANDERTHALS WHO FRUSTRATED EARLY MODERN HUMANS' FIRST ATTEMPTS TO MIGRATE OUT OF AFRICA.
ARCHAEOLOGICAL EVIDENCE SUGGESTS THAT THEY WERE VERY EFFECTIVE HUNTERS OF BIG GAME.
SO HOW WAS IT THAT MODERN HUMANS EVENTUALLY SUCCEEDED IN OVERCOMING THE NEANDERTHALS, SUPPLANTING BOTH THEM AND HOMO ERECTUS ACROSS THE EARTH?
YEAH-- HOW THAT HAPPEN?
ONE POSSIBLE ANSWER LIES IN FOXP2, A RECENTLY DISCOVERED GENE, THE HUMAN VERSION OF WHICH HELPS CONTROL SOPHISTICATED MUSCLE MOVEMENTS IN THE LOWER FACE AND LARYNX, AS WELL AS AIDING IN SPEECH RECOGNITION.
IT WAS PERHAPS THE FINAL REFINEMENT THAT ALLOWED THE DEVELOPMENT OF THE COMPLEXITIES OF HUMAN LANGUAGE.
A... E... I... O... U...
?
FOXP2 WAS DISCOVERED THROUGH THE EXTREME LINGUISTIC DISABILITIES EXPERIENCED BY MEMBERS OF A CONTEMPORARY ENGLISH FAMILY.
GENETICISTS FOUND THAT ALL AFFECTED FAMILY MEMBERS HAD INHERITED A DISABLED FOXP2 GENE FROM A COMMON GRANDMOTHER.
FOX GENES ARE ANCIENT-- MANY MAMMAL GENOMES, FROM MICE TO GORILLAS AND CHIMPANZEES, POSSESS VERSIONS.
BUT FOX GENES REMAINED STABLE, CONSERVATIVE GENES UNTIL THE HUMAN VERSION BEGAN AN ACCELERATED EVOLUTION...
HOOT.
WOOF.
MEOW.
SQUEAK.
...APPARENTLY NOT LONG AFTER THE CHIMP-HUMAN FORK IN THE PRIMATE FAMILY TREE OCCURRED. FOX GENES IN THE HUMAN GENOME CONTINUED THEIR MUTATIONS, WITH THE CURRENT VERSION DEVELOPING AS RECENTLY AS 200,000 YEARS AGO.
SOMEDAY OUR UNPARALLELED ORATORY SKILLS WILL WIN HEARTS AND MINDS AND SELL MANY USED CARS!

THE RAPIDLY EVOLVING VARIATION OF THE EARLY HUMAN FOX GENES WAS PERHAPS THE KEY FACTOR THAT ALLOWED THEM TO MASTER COMMUNICATION SKILLS FAR EXCEEDING THOSE OF ALL OTHER SPECIES.
NOW, TO RETURN TO THE EARLY HUMAN POPULATION'S NEANDERTHAL PROBLEM...
SOMETIME AROUND 50,000 YEARS AGO, A SMALL BAND OF MODERN HUMANS FINALLY MANAGED TO MOVE OUT OF AFRICA AND SURVIVE WITHIN THE MIDST OF AN ESTABLISHED NEANDERTHAL POPULATION.
IN AN UNDOUBTEDLY RELATED EVENT, HOMO NEANDERTHALENSIS AND HOMO ERECTUS, MODERN HUMANS' CLOSEST RELATIVES, DISAPPEARED COMPLETELY WITHIN THE NEXT 20,000 YEARS!
THAT'S A RELATIVELY SHORT PERIOD OF TIME FOR TWO PREVIOUSLY ROBUST SPECIES TO VANISH. MODERN HUMANS' MIGRATIONAL SUCCESSES THROUGH THIS SAME TIME FRAME WOULD SEEM TO INDICATE SOMETHING MORE THAN COINCIDENCE AT WORK.
WHETHER THESE SPECIES' DISAPPEARANCES WERE SINISTER OR NOT MAY BE DISCOVERED THROUGH FURTHER GENETIC INVESTIGATION.
THERE ARE TWO PREDOMINANT THEORIES CONCERNING THE NEANDERTHAL.
THEORY ONE
NEANDERTHALS WERE OUT-COMPETED BY MODERN HUMANS IN A BLOODY EXPANSIONIST EXTERMINATION. AS MODERN MAN GRADUALLY DISPLACED NEANDERTHALS ACROSS THE NORTHERN HEMISPHERE, THEY DESTROYED THEIR RIVALS, POPULATION BY POPULATION.
NEANDERTHAL REMAINS INDICATE THAT THEY WERE, INDIVIDUALLY, THE PHYSICALLY SUPERIOR SPECIES...
...BUT IF THEY DID NOT POSSESS THE SOPHISTICATED SOCIAL STRUCTURE AND LINGUISTIC SKILLS RECENTLY DEVELOPED BY MODERN HUMANS, THEY MAY HAVE REACHED A POINT OF SIGNIFICANT DISADVANTAGE.
?! NEW HUMAN SPEAR POINTS KILL BETTER AND BETTER...
AND HUMAN TEACH OTHER HUMAN TO MAKE NEW SPEAR SO QUICKLY!
THEORY TWO
THE NEANDERTHAL GENOME HAS BEEN ABSORBED INTO THE MODERN HUMAN GENOME THROUGH INTERBREEDING.
MAYBE NEANDERTHALS DID HAVE LINGUISTIC AND SOCIAL SKILLS THAT MODERN HUMANS FOUND ATTRACTIVE...
...AND VICE VERSA.

GENETIC ANTHROPOLOGISTS HOPE TO SOLVE THE MYSTERY WITH DNA THAT HAS BEEN RECENTLY EXTRACTED FROM A 38,000-YEAR-OLD NEANDERTHAL SKELETON.
OF PARTICULAR INTEREST AGAIN IS THE FOXP2 GENE.
IF THE NEANDERTHAL VERSION RESEMBLES THAT OF MODERN HUMANS, CHANCES ARE GREATLY INCREASED THAT NEANDERTHALS HAD COMPARABLE LINGUISTIC ABILITIES...
...CANCELING LANGUAGE AS A DISTINCT ADVANTAGE FOR THE MODERN HUMANS.
LOOK, WE COULD HURL DISPARAGING COMMENTS AT EACH OTHER ALL DAY OR WE COULD TRY AND GET BEYOND OUR DIFFERENCES.
OR, THE ULTIMATE MAPPING OF THE NEANDERTHAL GENOME COULD REVEAL MUCH INTERMIXING BETWEEN SPECIES— THE MODERN HUMAN GENOME MAY HAVE GOTTEN AS MUCH AS 25% OF ITS DNA FROM NEANDERTHALS...
...INDICATING THAT NEANDERTHALS DID NOT VANISH AS THE RESULT OF VIOLENT COMPETITION...
...BUT WERE, THROUGH THE HAPPIER CIRCUMSTANCES OF INTERBREEDING, INSTEAD LOST AS A SEPARATE SPECIES AS THEIR GENOME WAS INCORPORATED INTO THE HUMANS'!
MAYBE TO BE HUMAN IS TO BE NEANDERTHAL!
WHATEVER THE CIRCUMSTANCES OF THEIR RIVALS' DISAPPEARANCE, MODERN HUMANS DID SUCCEED IN THE INCREDIBLY DIFFICULT TASK OF BREAKING OUT OF THE ANCESTRAL AFRICAN HOMELAND...
...AND SUCCESSFULLY SPREAD ACROSS THE REST OF THE PLANET!
A GOOD GUESS WOULD BE THAT HUMAN POPULATION DRIFT ACROSS THE CONTINENTS WAS PROBABLY PROPELLED BY PRESSURE FROM WITHIN.
WHEN A LOCATION BECAME STRESSED BY OVERPOPULATION, A SEGMENT OF THAT POPULATION WAS FORCED TO MOVE FORWARD TO NEW TERRITORIES.
GO FIND YOUR OWN PLACE!

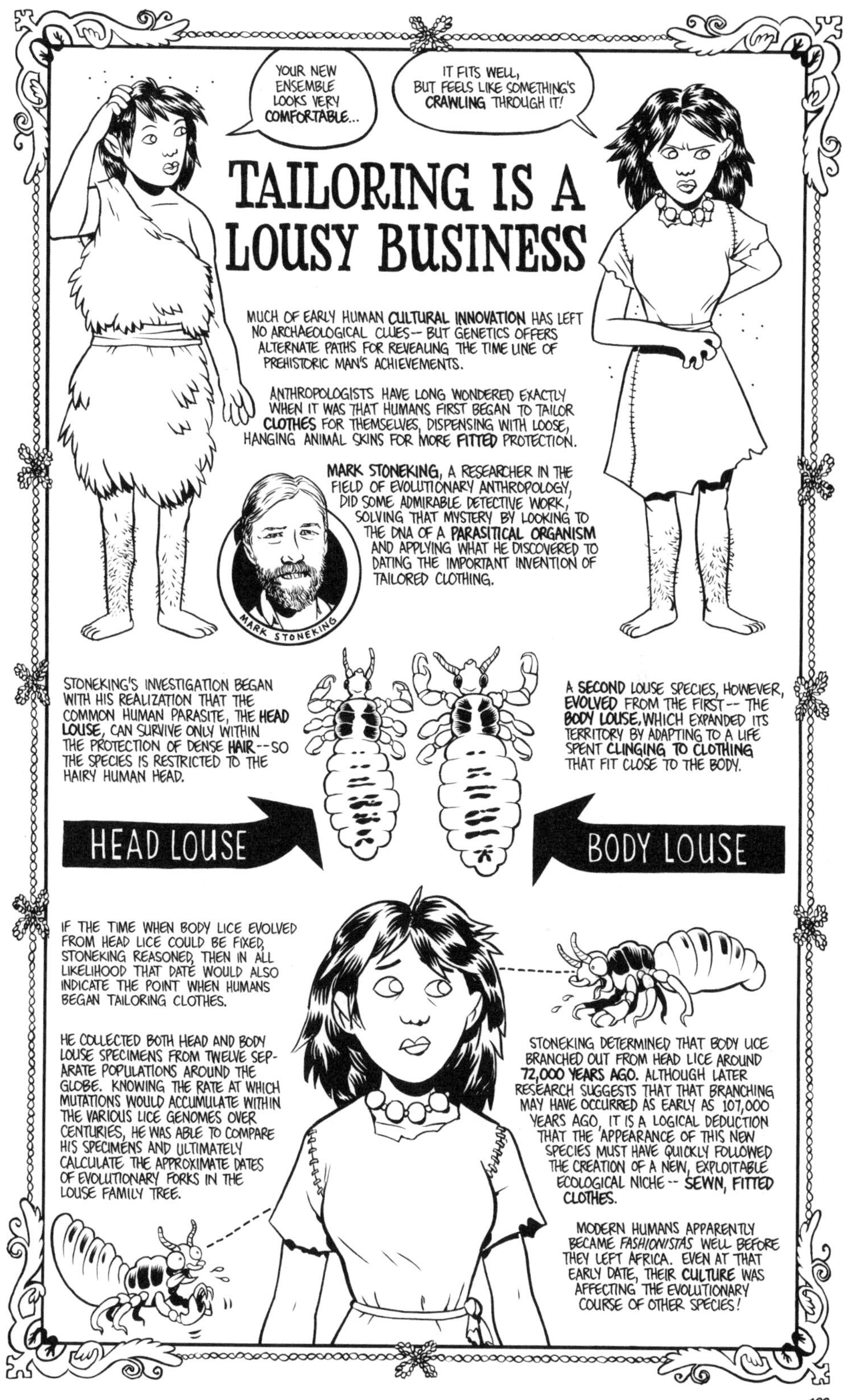
YOUR NEW ENSEMBLE LOOKS VERY COMFORTABLE...
IT FITS WELL, BUT FEELS LIKE SOMETHING'S CRAWLING THROUGH IT!
TAILORING IS A LOUSY BUSINESS
MUCH OF EARLY HUMAN CULTURAL INNOVATION HAS LEFT NO ARCHAEOLOGICAL CLUES-- BUT GENETICS OFFERS ALTERNATE PATHS FOR REVEALING THE TIME LINE OF PREHISTORIC MAN'S ACHIEVEMENTS.
ANTHROPOLOGISTS HAVE LONG WONDERED EXACTLY WHEN IT WAS THAT HUMANS FIRST BEGAN TO TAILOR CLOTHES FOR THEMSELVES, DISPENSING WITH LOOSE, HANGING ANIMAL SKINS FOR MORE FITTED PROTECTION.
MARK STONEKING, A RESEARCHER IN THE FIELD OF EVOLUTIONARY ANTHROPOLOGY, DID SOME ADMIRABLE DETECTIVE WORK, SOLVING THAT MYSTERY BY LOOKING TO THE DNA OF A PARASITICAL ORGANISM AND APPLYING WHAT HE DISCOVERED TO DATING THE IMPORTANT INVENTION OF TAILORED CLOTHING.
MARK STONEKING
STONEKING'S INVESTIGATION BEGAN WITH HIS REALIZATION THAT THE COMMON HUMAN PARASITE, THE HEAD LOUSE, CAN SURVIVE ONLY WITHIN THE PROTECTION OF DENSE HAIR--SO THE SPECIES IS RESTRICTED TO THE HAIRY HUMAN HEAD.
A SECOND LOUSE SPECIES, HOWEVER, EVOLVED FROM THE FIRST-- THE BODY LOUSE, WHICH EXPANDED ITS TERRITORY BY ADAPTING TO A LIFE SPENT CLINGING TO CLOTHING THAT FIT CLOSE TO THE BODY.
HEAD LOUSE
BODY LOUSE
IF THE TIME WHEN BODY LICE EVOLVED FROM HEAD LICE COULD BE FIXED, STONEKING REASONED, THEN IN ALL LIKELIHOOD THAT DATE WOULD ALSO INDICATE THE POINT WHEN HUMANS BEGAN TAILORING CLOTHES.
HE COLLECTED BOTH HEAD AND BODY LOUSE SPECIMENS FROM TWELVE SEPARATE POPULATIONS AROUND THE GLOBE. KNOWING THE RATE AT WHICH MUTATIONS WOULD ACCUMULATE WITHIN THE VARIOUS LICE GENOMES OVER CENTURIES, HE WAS ABLE TO COMPARE HIS SPECIMENS AND ULTIMATELY CALCULATE THE APPROXIMATE DATES OF EVOLUTIONARY FORKS IN THE LOUSE FAMILY TREE.
STONEKING DETERMINED THAT BODY LICE BRANCHED OUT FROM HEAD LICE AROUND 72,000 YEARS AGO. ALTHOUGH LATER RESEARCH SUGGESTS THAT THAT BRANCHING MAY HAVE OCCURRED AS EARLY AS 107,000 YEARS AGO, IT IS A LOGICAL DEDUCTION THAT THE APPEARANCE OF THIS NEW SPECIES MUST HAVE QUICKLY FOLLOWED THE CREATION OF A NEW, EXPLOITABLE ECOLOGICAL NICHE -- SEWN, FITTED CLOTHES.
MODERN HUMANS APPARENTLY BECAME FASHIONISTAS WELL BEFORE THEY LEFT AFRICA. EVEN AT THAT EARLY DATE, THEIR CULTURE WAS AFFECTING THE EVOLUTIONARY COURSE OF OTHER SPECIES!

MOST INDIVIDUALS POPULATING A NEW TERRITORY, HOWEVER, TENDED TO STAY IN THAT PLACE WHERE THEY HAD SET DOWN ROOTS, AND SO DID THE VAST MAJORITY OF THEIR DESCENDANTS.
GENERATIONS OF STABLE, ISOLATED POPULATIONS EVOLVED THEIR OWN DISTINCT VARIATIONS ON THE HUMAN GENOME.
THE TALL TREES BEAR DELICIOUS FRUIT, BUT ONLY I CAN REACH THEM!
EACH POPULATION'S GENOME WAS DEFINED THROUGH REGIONAL ENVIRONMENTAL PRESSURES, RANDOM MUTATION, AND NATURAL SELECTION.
WHEN THE EXCHANGE OF MUTATIONS--ALLELES--IS LIMITED TO A SPECIFIC POPULATION, THAT POPULATION DEVELOPS "SIGNATURE" GENETIC TRAITS, IDIOSYNCRATIC TRAITS PARTICULAR TO THAT POPULATION.
GEOGRAPHIC ISOLATION AND THE TENDENCY TO "STAY HOME" REINFORCE THOSE TRAITS WITHIN THE POPULATION.
WHY, YES, I DO COME FROM THE LAND OF THE TALL TREES...
ENVIRONMENTAL FACTORS ARE PERHAPS MOST OFTEN THE FORCES BEHIND GENOME VARIATION WITHIN POPULATIONS.
FOR INSTANCE...
...DARK SKIN AND HAIR OFFER THE GREATEST PROTECTION AGAINST SOLAR RADIATIONS AND ARE FOUND MOST COMMONLY IN POPULATIONS LOCATED CLOSEST TO EARTH'S EQUATOR, WHERE THE SUN'S RAYS ARE THE MOST DIRECT...
...WHILE IN REGIONS CLOSER TO THE POLES, WHERE SUNLIGHT IS AT A PREMIUM, GENOMES FEATURING PALE SKIN AND HAIR, WHICH ABSORB SUNLIGHT MORE READILY, ARE MORE COMMON.
A MORE SPECIFIC AND COMPLEX EXAMPLE OF POPULATION GENETICS WOULD BE THAT OF LACTOSE TOLERANCE--
--THE ABILITY TO DIGEST MILK.

AS HUMAN SOCIETIES DIVERSIFIED -- PARTIALLY AS A RESULT OF THE ABILITY TO COMMUNICATE WITH SOPHISTICATION -- SOME POPULATIONS GAVE UP THE HUNTER-GATHERER NOMADIC EXISTENCE IN FAVOR OF SETTLED AGRICULTURE.
PEOPLES IN EUROPE AND AFRICA WHO DOMESTICATED CATTLE FOUND IT AN ADVANTAGE TO INCORPORATE THEIR STOCK'S MILK INTO THEIR DIET.
MILK IS FULL OF CALORIES -- MOST OF WHICH COME FROM THE SUGAR LACTOSE. THE ENZYME NECESSARY FOR DIGESTING LACTOSE IS LACTASE, WHICH IS CODED BY A GENE THAT IS GENERALLY SWITCHED OFF WHEN AN INDIVIDUAL IS WEANED.
MILK IS IMPORTANT FOR INFANT GROWTH BUT IS NOT A RESOURCE NATURALLY AVAILABLE TO ADULTS, SO THE PRODUCTION OF LACTASE BECOMES EXPENDABLE.
CATTLE-HERDING PEOPLES IN AFRICA AND EUROPE DEVELOPED MUTATIONS THAT KEPT THE LACTASE PRODUCTION GENE SWITCHED PERMANENTLY ON. THEY PROSPERED WITH THEIR MILK-SUPPLEMENTED DIET AND SO HAD A BETTER CHANCE OF REPRODUCING AND PASSING ON THEIR NEW ALLELE TO THEIR OFFSPRING.
OF COURSE, IF THE LACTOSE-TOLERANCE ALLELE APPEARED IN POPULATIONS THAT WERE NOT HERDERS, IT WAS OF NO ADVANTAGE.
ANOTHER EXAMPLE OF POPULATION GENETICS ENTAILS PROTECTIONS AGAINST MALARIA, A TROPICAL DISEASE THOUGHT TO HAVE BECOME A DEADLY THREAT IN RELATIVELY RECENT HUMAN HISTORY...
... AS HUMAN AGRICULTURAL CULTIVATION OPENED UP VAST BREEDING HABITATS FOR MOSQUITOES CARRYING THE MALARIA PARASITE.
MUTATIONS THAT ALTER HEMOGLOBIN AND PROVIDE A PROTECTION AGAINST MALARIA WERE SELECTED IN INDEPENDENT POPULATIONS IN AFRICA AND IN POPULATIONS BORDERING THE MEDITERRANEAN SEA.
THEY SAY THE BLOOD TASTED DIFFERENT BACK IN GREAT-GRANDDAD'S DAY.
IN A CASE OF UNFORTUNATE PLEIOTROPY, HOWEVER...
... THOSE NEW ALLELES RESPONSIBLE FOR HINDERING MALARIA ALSO GAVE RISE TO VARIOUS BLOOD DISEASES...
... MOST NOTORIOUSLY, SICKLE-CELL ANEMIA.

THE APPEARANCE OF BOTH LACTOSE TOLERANCE AND MALARIA RESISTANCE ILLUSTRATES SOME IMPORTANT ASPECTS OF HUMAN EVOLUTION:
1) THE HUMAN GENOME IS NOT A FIXED THING -- IT IS CONSTANTLY CHANGING.
2) WIDELY SEPARATED POPULATIONS, RESPONDING TO SIMILAR CONDITIONS, CAN INDEPENDENTLY ARRIVE AT SIMILAR GENETIC SOLUTIONS.
YOU MEAN TO SAY THAT YOU GOT MILK, TOO?
3) GENES CAN RESPOND TO ENVIRONMENTAL CONDITIONS, INCLUDING CULTURAL AND BEHAVIORAL FACTORS.
SEEMS THAT THERE ARE MORE AND MORE PESKY MOSQUITOES FOR EVERY HECTARE WE CLEAR!
YEAH, BUT AT LEAST WE'RE NOT GETTING AS SICK AS OUR FATHERS' GENERATION DID...
OF COURSE, THE GENOMES OF ALL OTHER ORGANISMS ARE LIKEWISE CONSTANTLY MUTATING, ADAPTING TO CHANGING ENVIRONMENTAL CONDITIONS -- INCLUDING THOSE FOR WHICH HUMANS ARE RESPONSIBLE.
SWAT!!
EVERY SPECIES SUCCESSFULLY EVOLVES -- OR FAILS -- IN RESPONSE TO ALL THE OTHER SPECIES WITH WHICH IT COMES INTO CONTACT!
I DON'T KNOW HOW WE MANAGED TO SURVIVE BEFORE THESE MAMMALIAN BIPEDS CAME ALONG!
YIKES! What you've just covered gives me HOPE --
-- maybe somewhere here on Glargal there are SQUINCH with mutations that are even now countering our genetic disorder...
THAT'S A POSSIBILITY, ALTHOUGH WITHOUT SEXUAL REPRODUCTION, THE DISPERSION OF ANY HELPFUL MUTATION WOULD BE AWFULLY SLOW.
BUT YOUR THOUGHTS ARE HEADED IN THE RIGHT DIRECTION, MOST SAGACIOUS SIRE!
ON EARTH, THE HUMANS ARE NOW LOOKING AT THE GENETIC IDIOSYNCRASIES OF SPECIFIC POPULATIONS TO DETERMINE WHAT CAN BE LEARNED OF EACH POPULATION'S SUCCESSES AND LIABILITIES REGARDING DISEASES AND DISORDERS.

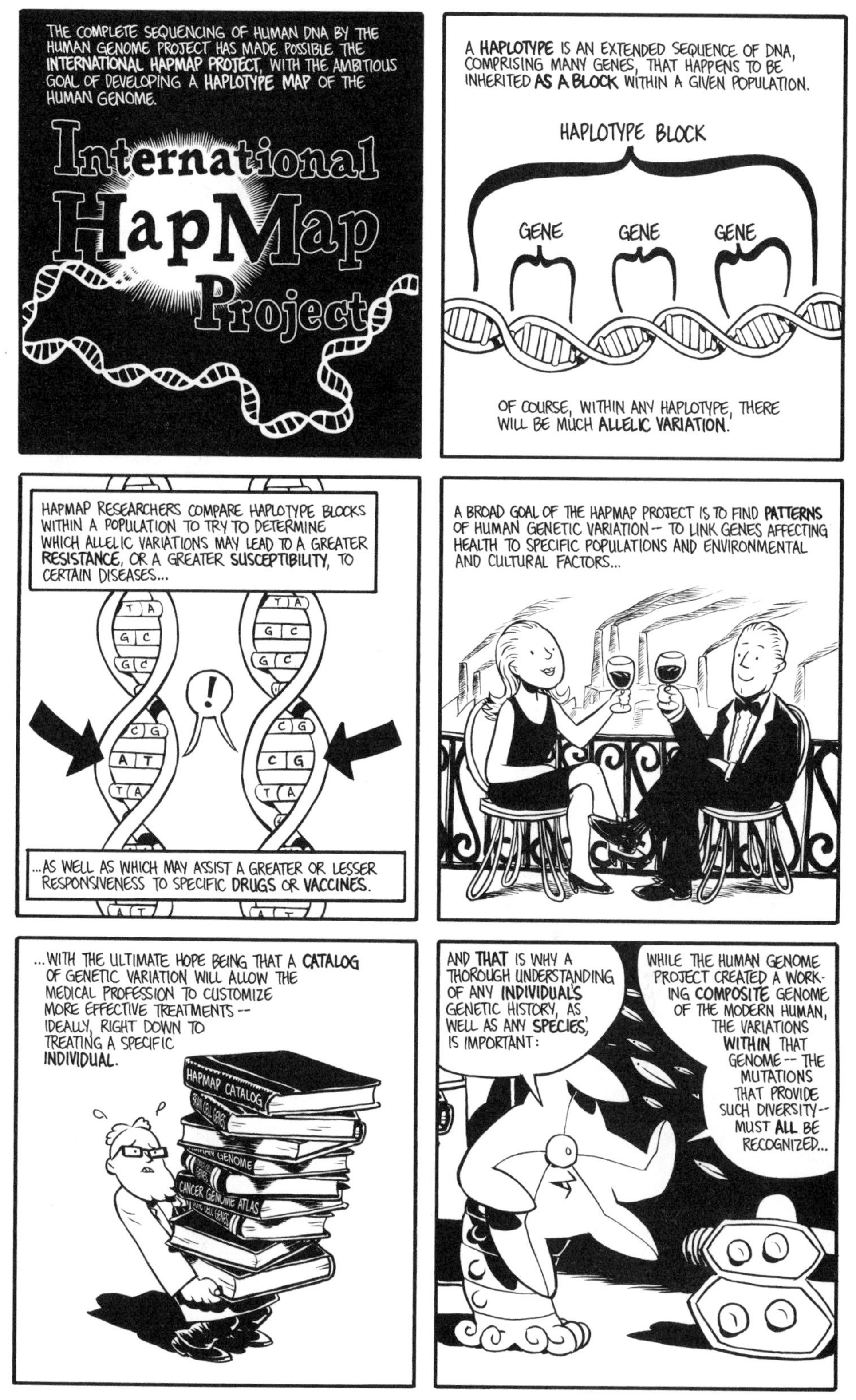
THE COMPLETE SEQUENCING OF HUMAN DNA BY THE HUMAN GENOME PROJECT HAS MADE POSSIBLE THE **INTERNATIONAL HAPMAP PROJECT**, WITH THE AMBITIOUS GOAL OF DEVELOPING A **HAPLOTYPE MAP** OF THE HUMAN GENOME.
International HapMap Project
A **HAPLOTYPE** IS AN EXTENDED SEQUENCE OF DNA, COMPRISING MANY GENES, THAT HAPPENS TO BE INHERITED **AS A BLOCK** WITHIN A GIVEN POPULATION.
HAPLOTYPE BLOCK
GENE
GENE
GENE
OF COURSE, WITHIN ANY HAPLOTYPE, THERE WILL BE MUCH **ALLELIC VARIATION**.
HAPMAP RESEARCHERS COMPARE HAPLOTYPE BLOCKS WITHIN A POPULATION TO TRY TO DETERMINE WHICH ALLELIC VARIATIONS MAY LEAD TO A GREATER **RESISTANCE**, OR A GREATER **SUSCEPTIBILITY**, TO CERTAIN DISEASES...
!
...AS WELL AS WHICH MAY ASSIST A GREATER OR LESSER RESPONSIVENESS TO SPECIFIC **DRUGS** OR **VACCINES**.
A BROAD GOAL OF THE HAPMAP PROJECT IS TO FIND **PATTERNS** OF HUMAN GENETIC VARIATION -- TO LINK GENES AFFECTING HEALTH TO SPECIFIC POPULATIONS AND ENVIRONMENTAL AND CULTURAL FACTORS...
...WITH THE ULTIMATE HOPE BEING THAT A **CATALOG** OF GENETIC VARIATION WILL ALLOW THE MEDICAL PROFESSION TO CUSTOMIZE MORE EFFECTIVE TREATMENTS -- IDEALLY, RIGHT DOWN TO TREATING A SPECIFIC **INDIVIDUAL**.
HAPMAP CATALOG
CANCER GENOME ATLAS
AND **THAT** IS WHY A THOROUGH UNDERSTANDING OF ANY **INDIVIDUAL'S** GENETIC HISTORY, AS WELL AS ANY **SPECIES'**, IS IMPORTANT:
WHILE THE HUMAN GENOME PROJECT CREATED A WORKING **COMPOSITE** GENOME OF THE MODERN HUMAN, THE VARIATIONS **WITHIN** THAT GENOME -- THE MUTATIONS THAT PROVIDE SUCH DIVERSITY -- MUST **ALL** BE RECOGNIZED...

...IF THE MOST EFFECTIVE REMEDIES POSSIBLE -- THOSE DESIGNED EXPRESSLY FOR THE INDIVIDUAL, WHETHER TO TREAT MALADIES THAT ARE DIRECTLY THE RESULT OF GENETIC DISORDERS...
1 2 3 4 5
6 7 8 9 10 11
12 13 14 15 16 17
18 19 20 21 22
DOWN SYNDROME
NORMAL RED BLOOD CELLS
SICKLE-CELL-ANEMIA-AFFLICTED RED BLOOD CELLS
...OR FOR THOSE MADE MORE LIKELY BY GENETIC VARIATIONS -- ARE TO BECOME A REALITY.
VERY RECENTLY, JAMES WATSON -- ONE MEMBER OF THE TEAM WHO DISCOVERED THE MOLECULAR STRUCTURE OF DNA -- HAD HIS PERSONAL GENOME SEQUENCED...
...AT A COST OF UNDER $1 MILLION.
THAT'S A SIGNIFICANT MONETARY FIGURE, IF SOMEWHAT SYMBOLIC. PREVIOUSLY, IT HAD COST MANY MILLIONS OF DOLLARS TO SEQUENCE A GENOME.
GENOME SEQUENCING now only $999,999.99!
BUT CONTINUING BREAKTHROUGHS IN SEQUENCING TECHNOLOGIES PROMISE THE FURTHER REDUCTION OF THAT EXPENSE.
THERE IS HOPE THAT THE COST OF PERSONAL GENOME SEQUENCING CAN BE REDUCED TO A FEW THOUSAND DOLLARS AND SO BECOME AVAILABLE TO ALL HUMAN PATIENTS...
...PROVIDING VERY SPECIFIC MAPS FOR THE CARE OF THE INDIVIDUAL...
...AS OPPOSED TO THE VERY GENERALIZED CARE NOW AVAILABLE.

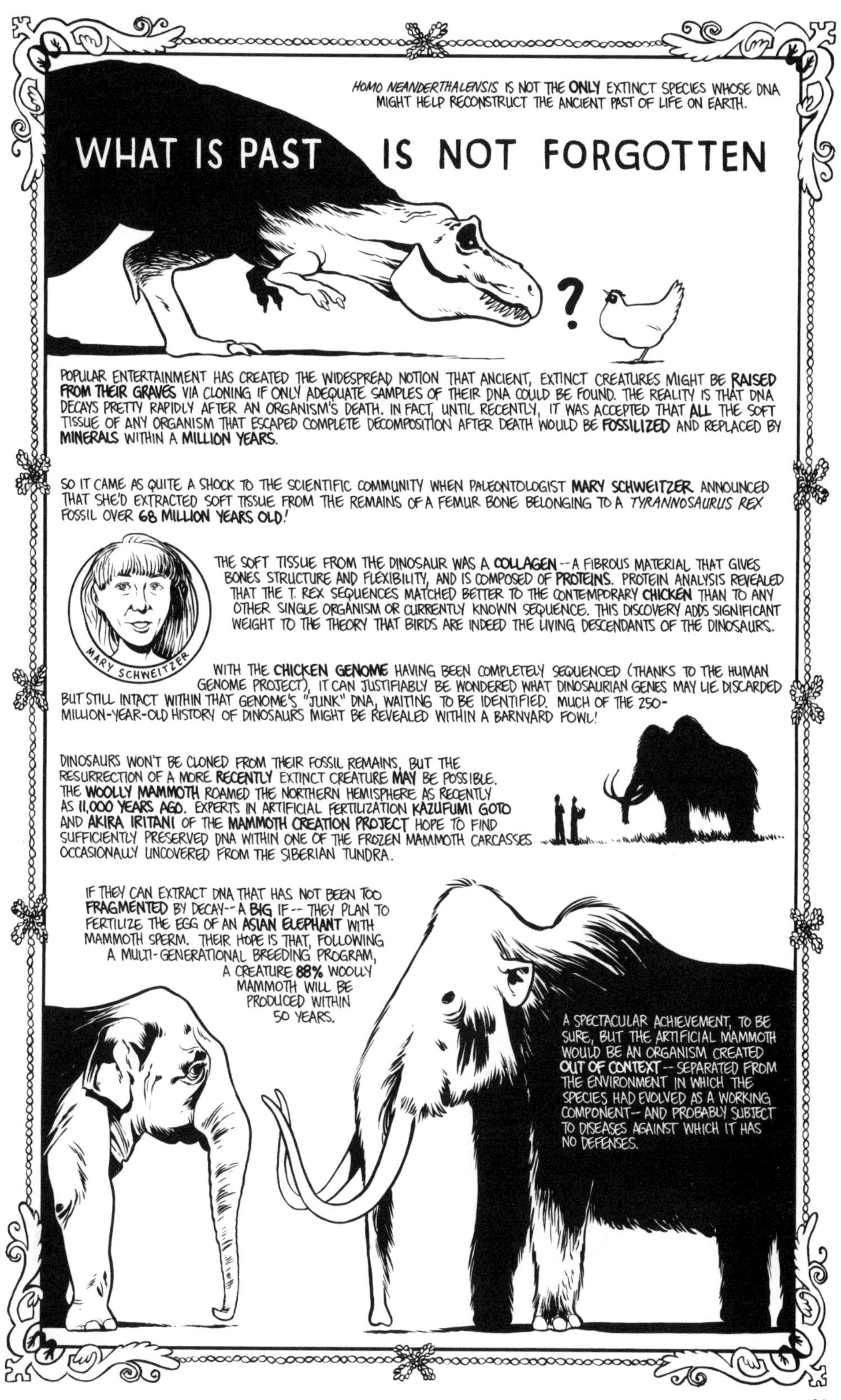
HOMO NEANDERTHALENSIS IS NOT THE ONLY EXTINCT SPECIES WHOSE DNA MIGHT HELP RECONSTRUCT THE ANCIENT PAST OF LIFE ON EARTH.
WHAT IS PAST IS NOT FORGOTTEN
?
POPULAR ENTERTAINMENT HAS CREATED THE WIDESPREAD NOTION THAT ANCIENT, EXTINCT CREATURES MIGHT BE RAISED FROM THEIR GRAVES VIA CLONING IF ONLY ADEQUATE SAMPLES OF THEIR DNA COULD BE FOUND. THE REALITY IS THAT DNA DECAYS PRETTY RAPIDLY AFTER AN ORGANISM'S DEATH. IN FACT, UNTIL RECENTLY, IT WAS ACCEPTED THAT ALL THE SOFT TISSUE OF ANY ORGANISM THAT ESCAPED COMPLETE DECOMPOSITION AFTER DEATH WOULD BE FOSSILIZED AND REPLACED BY MINERALS WITHIN A MILLION YEARS.
SO IT CAME AS QUITE A SHOCK TO THE SCIENTIFIC COMMUNITY WHEN PALEONTOLOGIST MARY SCHWEITZER ANNOUNCED THAT SHE'D EXTRACTED SOFT TISSUE FROM THE REMAINS OF A FEMUR BONE BELONGING TO A TYRANNOSAURUS REX FOSSIL OVER 68 MILLION YEARS OLD!
MARY SCHWEITZER
THE SOFT TISSUE FROM THE DINOSAUR WAS A COLLAGEN--A FIBROUS MATERIAL THAT GIVES BONES STRUCTURE AND FLEXIBILITY, AND IS COMPOSED OF PROTEINS. PROTEIN ANALYSIS REVEALED THAT THE T. REX SEQUENCES MATCHED BETTER TO THE CONTEMPORARY CHICKEN THAN TO ANY OTHER SINGLE ORGANISM OR CURRENTLY KNOWN SEQUENCE. THIS DISCOVERY ADDS SIGNIFICANT WEIGHT TO THE THEORY THAT BIRDS ARE INDEED THE LIVING DESCENDANTS OF THE DINOSAURS.
WITH THE CHICKEN GENOME HAVING BEEN COMPLETELY SEQUENCED (THANKS TO THE HUMAN GENOME PROJECT), IT CAN JUSTIFIABLY BE WONDERED WHAT DINOSAURIAN GENES MAY LIE DISCARDED BUT STILL INTACT WITHIN THAT GENOME'S "JUNK" DNA, WAITING TO BE IDENTIFIED. MUCH OF THE 250-MILLION-YEAR-OLD HISTORY OF DINOSAURS MIGHT BE REVEALED WITHIN A BARNYARD FOWL!
DINOSAURS WON'T BE CLONED FROM THEIR FOSSIL REMAINS, BUT THE RESURRECTION OF A MORE RECENTLY EXTINCT CREATURE MAY BE POSSIBLE. THE WOOLLY MAMMOTH ROAMED THE NORTHERN HEMISPHERE AS RECENTLY AS 11,000 YEARS AGO. EXPERTS IN ARTIFICIAL FERTILIZATION KAZUFUMI GOTO AND AKIRA IRITANI OF THE MAMMOTH CREATION PROJECT HOPE TO FIND SUFFICIENTLY PRESERVED DNA WITHIN ONE OF THE FROZEN MAMMOTH CARCASSES OCCASIONALLY UNCOVERED FROM THE SIBERIAN TUNDRA.
IF THEY CAN EXTRACT DNA THAT HAS NOT BEEN TOO FRAGMENTED BY DECAY-- A BIG IF-- THEY PLAN TO FERTILIZE THE EGG OF AN ASIAN ELEPHANT WITH MAMMOTH SPERM. THEIR HOPE IS THAT, FOLLOWING A MULTI-GENERATIONAL BREEDING PROGRAM, A CREATURE 88% WOOLLY MAMMOTH WILL BE PRODUCED WITHIN 50 YEARS.
A SPECTACULAR ACHIEVEMENT, TO BE SURE, BUT THE ARTIFICIAL MAMMOTH WOULD BE AN ORGANISM CREATED OUT OF CONTEXT-- SEPARATED FROM THE ENVIRONMENT IN WHICH THE SPECIES HAD EVOLVED AS A WORKING COMPONENT-- AND PROBABLY SUBJECT TO DISEASES AGAINST WHICH IT HAS NO DEFENSES.

AND THAT...
...UMM...
...THAT CONCLUDES MY REPORT ON GENETICS FROM THE PLANET EARTH...
Hmm.
As you chose to end by emphasizing the importance that humans place on understanding genetics on an individual basis...
...I assume you are suggesting that this is the path WE follow?
ONLY --GULP-- ONLY...
...ONLY BECAUSE IT IS IN THE ROYAL FAMILY'S BEST INTERESTS TO HAVE AS MANY GENETIC OPTIONS AVAILABLE AS POSSIBLE.
IF THE REST OF SQUINCH-DOM ALSO HAPPENS TO BENEFIT, SO MUCH THE BETTER.
IF OUR GENETIC ORGANIZATION IS ANYTHING LIKE THAT ON EARTH, THE MORE WE LEARN ABOUT EACH OTHER...
... THE MORE WE WILL HELP OURSELVES!
AND THE MORE WE UNDERSTAND OF OUR EVOLUTIONARY HISTORY, THE BETTER GRIP WE'LL HAVE ON OUR CURRENT CRISIS...
Okay, I GET it.

WE MUST--
Okay, OKAY! You've SOLD me, Bloort!
LISTEN UP, EVERYBODY!
I'm making a GLOBAL DECLARATION OF INTENT, by the power vested in me as LORD OF ALL SQUINCHDOM!
On this day, we launch a TOTAL SCIENTIFIC COMMITMENT to uncover our genetic heritage!
We'll learn all about our MOLECULAR, CELLULAR, and TRANSMISSION strategies!
We will even investigate the possibility of improving our genetic options by developing technologies that will allow us to -- ugh -- SWAP GENES.
AND, Bloort 183 -- we will research our GENETIC HISTORY...
... EVERY Squinch's genetic history...
...applying that knowledge to develop therapies that'll BEAT this disorder!
MIGHT I SUGGEST, YOUR MOST GRACIOUS OMNIPOTENCE, THAT MY UNIQUE EXPERIENCE WOULD MAKE ME THE PERFECT CHOICE TO HEAD UP OUR--
NONSENSE, Bloort! You're FAR too valuable on that fascinating planet Earth, collecting more information! I want you back there IMMEDIATELY!
LET'S GET TO WORK, FOLKS!
Where's my speechwriter?!

IT IS A FASCINATING PLANET.

Suggested Reading

Glossary

SUGGESTED READING

PERIODICALS

SCIENTIFIC AMERICAN (www.SciAm.com)

NATURAL HISTORY (www.naturalhistorymag.com)

NATIONAL GEOGRAPHIC (www.nationalgeographic.com)

DISCOVER (www.discovermagazine.com)

NEW SCIENTIST (www.newscientist.com)

SCIENCE NEWS (www.sciencenews.org)

BOOKS

Dawkins, Richard. *THE SELFISH GENE*, 3rd ed. New York: Oxford University Press, 2006.

Fortey, Richard. *LIFE: A NATURAL HISTORY OF THE FIRST FOUR BILLION YEARS OF LIFE ON EARTH.* New York: Vintage Books, 1999.

Fritz, Sandy, ed. *UNDERSTANDING CLONING: FROM THE EDITORS OF SCIENTIFIC AMERICAN.* New York: Warner Books, 2002.

Hartl, Daniel L., and Elizabeth W. Jones. *GENETICS: ANALYSIS OF GENES AND GENOMES*, 6th ed. Boston: Jones and Bartlett, 2004.

Hartwell, Leland, et al. *GENETICS: FROM GENES TO GENOMES*, 3rd ed. New York: McGraw-Hill, 2006.

Pollan, Michael. *THE BOTANY OF DESIRE.* New York: Random House, 2001.

Wade, Nicholas. *BEFORE THE DAWN: RECOVERING THE LOST HISTORY OF OUR ANCESTORS.* New York: Penguin, 2006.

Watson, James D., and Berry Andrew. *DNA: THE SECRET OF LIFE.* New York: Knopf, 2004.

WEBSITES

GENOME NEWS NETWORK
www.genomenewsnetwork.org

HOWARD HUGHES MEDICAL INSTITUTE
www.hhmi.org

HUMAN GENOME PROJECT INFORMATION
www.ornl.gov/sci/techresources/Human_Genome/home.shtml

NATIONAL CENTER FOR BIOTECHNOLOGY INFORMATION
www.ncbi.nlm.nih.gov

NATIONAL HUMAN GENOME RESEARCH INSTITUTE
www.genome.gov

PALOMAR COLLEGE (for specific information about Mendelian genetics)
anthro.palomar.edu/mendel

UNIVERSITY OF CALIFORNIA BERKELEY (for specific information about evolution)
evolution.berkeley.edu

GLOSSARY

ADENINE: Nucleic acid of both DNA and RNA.

ADENOVIRAL VECTORS: Transmission of an infectious agent to a host cell using the viral coat and molecular mechanisms of an Adenovirus.

ADENOVIRUSES: A group of DNA-containing viruses that cause upper respiratory tract infections in humans.

ALLELES: Gene variants with the same locus on homologous chromosomes.

AMINO ACIDS: Building blocks of proteins; they contain amino and carboxylic acid groups.

ANAPHASE: Phase of cell division during which sister chromatids separate into independent chromosomes and the cleavage furrow forms.

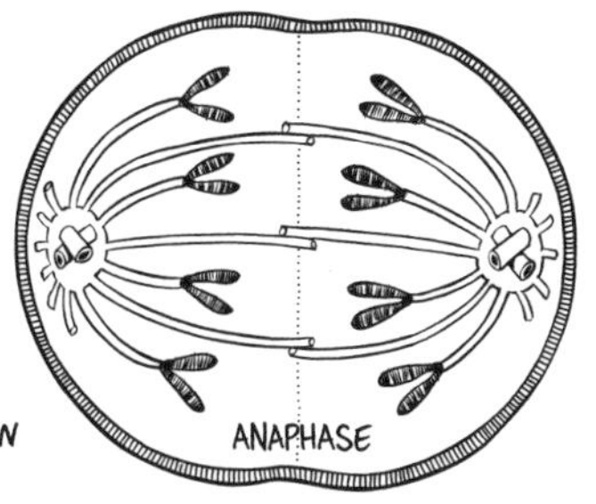

ANEUPLOIDY: An abnormal chromosome number that is not an exact multiple of the haploid number.

ANTI-CODON: A sequence of three adjacent nucleotides that bonds to a corresponding codon in messenger RNA (mRNA) during protein synthesis.

AUTOSOMAL DOMINANT: A phenotype resulting from the dominant expression of one allele over the other in a gene pair.

AUTOSOMAL RECESSIVE: A phenotype in which a homozygous recessive genotype is expressed.

AUTOSOMES: All chromosomes that do not serve to determine the sex of an individual organism.

AUTOTROPHS: Organisms that are able to independently synthesize organic molecules from inorganic material and/or convert solar energy into usable energy.

BACTERIA: A single-celled organism that lacks a defined nuclear membrane.

BRCA 1 GENE: Breast Cancer 1 Gene—a tumor suppressor gene, commonly mutated in females with breast and/or ovarian cancer, that normally serves to repair damaged DNA.

CANCER: A group of diseases similarly characterized by uncontrolled cellular replication.

CENTRIOLES: Organelles involved in spindle formation during the cell-division stage called anaphase.

CENTROMERE: The region at which chromatids are joined and to which spindle fibers attach during cell division.

CHROMATID: Either of two daughter strands of a duplicated chromosome that are joined by a centromere and separate during cell division to become individual chromosomes.

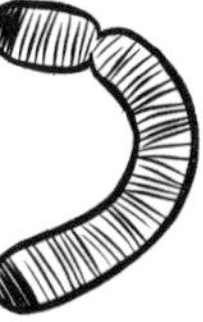

CHROMOSOME: A linear arrangement of tightly coiled DNA located in a cell's nucleus.

CLONING: Duplication of a biological entity.

CODOMINANCE: The result of nonidentical alleles at a heterozygous gene locus expressing fully as both phenotypes.

CODON: A specific sequence of three adjacent nucleic acids on a strand of DNA that code for an amino acid.

CRI DU CHAT SYNDROME: A congenital disease resulting from a partial deletion of chromosome 5, characterized by mental retardation and a catlike cry.

CYSTIC FIBROSIS: An autosomal recessive disease caused by a mutation in the cystic fibrosis transmembrane conductance regulator (CFTR) gene.

CYTOKINESIS: Separation of the cytoplasm of the parent cell as two daughter cells form during cell division.

CYTOPLASM (aka protoplasm): The fluid substance that fills the cell.

CYTOSINE: Nucleic acid of both DNA and RNA.

DIPLOID: Cells and/or organisms having two copies of each chromosome.

DNA: Deoxyribonucleic acid. It carries the genetic information in the cell and is capable of self-replication..

DNA

DOWN SYNDROME: A developmental disorder in humans characterized by mental retardation resulting from an extra copy of chromosome 21.

DUCHENNE MUSCULAR DYSTROPHY: An X-linked recessive disease caused by a defective gene for dystrophin and resulting in progressive muscle atrophy and weakness.

ENDOPLASMIC RETICULUM: A membrane network within the cytoplasm of cells involved in synthesis, modification, and transport of cellular material.

EPISTASIS: Suppression of one gene by another unrelated gene.

EUKARYOTE: An organism composed of cells containing a nucleus with linear chromosomes contained within a nuclear membrane, and which replicates by mitosis.

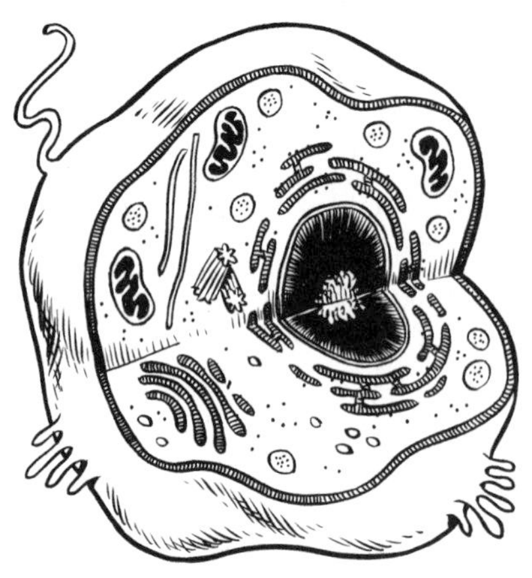

EUKARYOTE CELL

EXONS: Sequences of DNA that code for proteins.

FRAGILE X SYNDROME: An X-linked dominant disease caused by a mutation in the FMR1 gene resulting in mental retardation and autism.

FUNGUS: Any organism from the kingdom Fungi, including yeasts, molds, and mushrooms.

FUNGUS

GEMMULE: The bud of an asexually reproducing parent organism that becomes its offspring.

GENE: A heritable unit of DNA that encodes or directly contributes to protein synthesis.

GENETIC DRIFT: Reduction of heterozygosity within a population due to the random reduction of an allele in the gene pool over time; inversely related to the population size.

GENOME: The entire complement of genes in a species.

GENOTYPE: The entire genetic constitution of an individual.

GONOSOMES: Sex-determining chromosomes.

GUANINE: Nucleic acid of both DNA and RNA.

GYRASE: An enzyme that increases the density of the DNA helix by introducing negative supercoils and removing positive supercoils.

HAPLOID ORGANISM

HAPLOID: A cell or organism having only one copy of each chromosome.

HAPLOTYPE: A group of closely related genes often inherited as a unit.

HELICASE: An enzyme that aids in the separation of double-stranded DNA, thereby allowing single-stranded DNA to be copied.

HETEROTROPHS: Organisms that rely on other organisms for their energy requirements.

HETEROZYGOUS: Term applied to alleles in a gene pair that are different at a given chromosomal locus.

HISTONES: Proteins in a cell's nucleus around which DNA tightly coils.

HOLOENZYME COMPLEX: A complex of approximately 20 proteins, including transcription factors, that binds to the promoter region to initiate transcription.

HOMOLOGOUS: Similar in structure and evolutionary origin, though not necessarily in function.

HOMOZYGOUS: Term applied to identical alleles in the gene pair at a given chromosomal locus.

HUNTINGTON'S DISEASE: An autosomal dominant disease caused by a mutation in the Huntington (Htt) gene, leading to an alteration in the Htt protein and neuronal death.

INCOMPLETE DOMINANCE: Partial expression of non-identical alleles resulting in an intermediate phenotype.

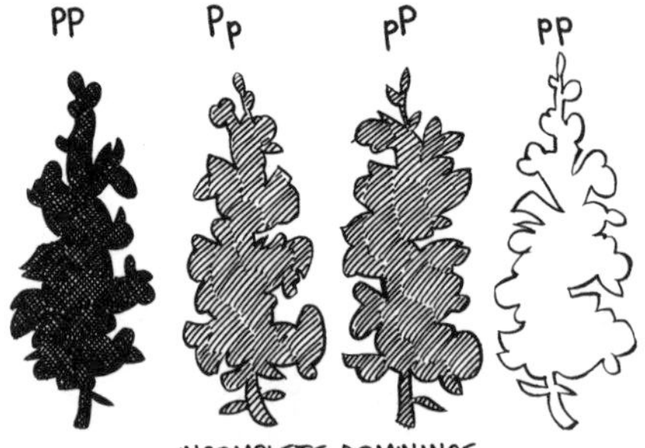

INCOMPLETE DOMINANCE

INCOMPLETE PENETRANCE: A condition, due to environmental conditions or the interactions of non-allelic genes, where the genotype is not fully realized in the phenotype.

INCONTINENTIA PIGMENTI: An X-linked dominant disease caused by mutations in the NEMO gene and resulting in multiple disorders of the skin and nervous system.

INDEPENDENT ASSORTMENT: A Mendelian law stating that maternal and paternal chromosomes are bequeathed to gametes independently of one another during meiosis.

INTRONS: Non-coding DNA that exists between exons.

KLINEFELTER'S SYNDROME: A chromosomal anomaly in males affecting sexual development and resulting in sterility.

LENTIVIRAL VECTORS: Means of transmission of an infectious agent to a host cell using the viral coat and molecular mechanisms of a lentivirus.

LENTIVIRUSES: A subclass of retrovirus that is able to infect non-dividing cells as well as dividing cells.

LIGASE: An enzyme that joins unattached ends of DNA.

LYSOSOME: A small cavity in the cell's cytoplasm containing enzymes involved in protein degradation and natural cell death.

MARFAN SYNDROME: An autosomal dominant disorder with multiple defects caused by mutations in the gene coding for fibrilin, a protein of the connective tissue.

MEIOSIS: The process by which gametes--male and female sex cells--are formed.

MESSENGER RNA (mRNA): RNA that mediates the transfer of genetic information from a cell nucleus to ribosomes in the cytoplasm, where it serves as a template for protein synthesis.

METAPHASE: The second stage of cell division in which the chromosomes are arranged in the equatorial plane prior to separation.

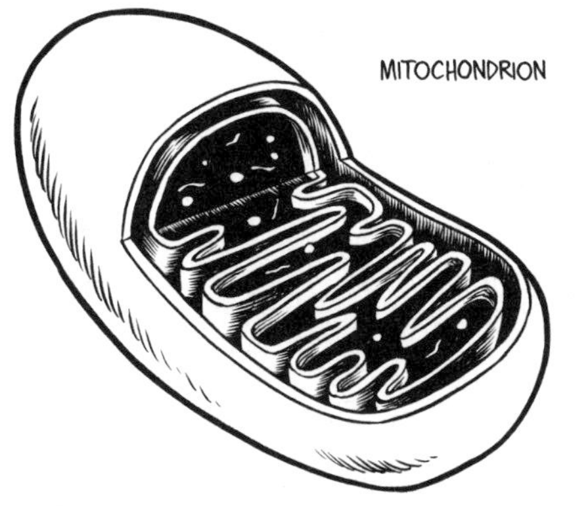

MITOCHONDRIA: Organelles in the cytoplasm of eukaryote cells important for cell metabolism.

MITOSIS: The process of cellular and genetic replication.

MUTATION: An inconsistency in an organism's genetic code.

NUCLEOSOME: A protein core of histones around which DNA tightly coils.

ONCORETROVIRAL VECTOR: Means of transmission of an infectious agent to a host cell using the viral coat and molecular mechanisms of an oncoretrovirus.

ONCORETROVIRUS: A retroviral virus commonly used as a vector for gene therapy.

PARKINSON'S DISEASE: A disease often caused by insufficient production and/or action of dopamine; characterized by progressive degeneration of the nervous system.

PARTHENOGENESIS: Development of an unfertilized gamete into an adult organism.

PHENOTYPE: Observed genetic expression of an organism's physical, biochemical, and physiological characteristics.

PLEIOTROPY: The expression of multiple phenotypes from a single gene.

POLYMERASE: An enzyme that aids in the attachment of nucleic acids to one another during the transcription process.

POLYPEPTIDE: A polymer of amino acids linked by peptide bonds; may be a protein fragment.

POLYSACCHARIDES: Carbohydrates composed of simple sugars.

PROGESTERONE: A hormone that acts to prepare the uterus for implantation of a fertilized ovum and helps maintain a healthy pregnancy.

PROKARYOTE: A single-celled organism that lacks a membrane-bound nucleus.

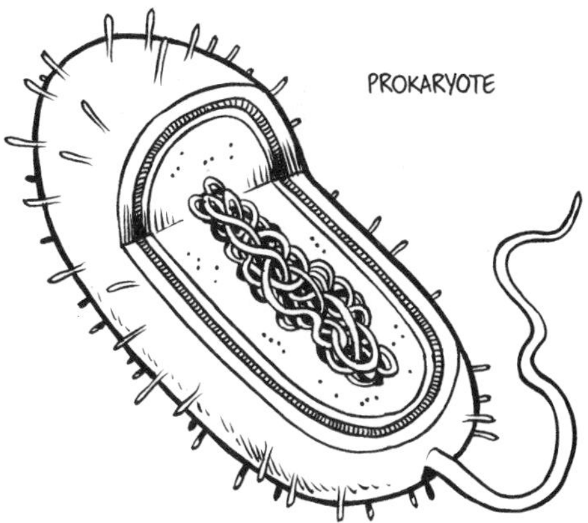

PROPHASE: In cell division, the stage where centrosomes move to opposite poles and spindle fibers form.

PROTEINS: Polymers of amino acids linked by peptide bonds.

PROTIST: Single-celled, eukaryotic organisms, including protozoans, slime molds, and certain algae.

RETROVIRUS: Any virus with an RNA genome that, following entry into a host cell, is reverse transcribed into double-stranded DNA and integrates into the host genome.

RETROVIRUS

RIBOSOMAL RNA (rRNA): A class of RNA found in the ribosomes of cells involved in protein synthesis.

RIBOSOMES: Organelles involved in protein synthesis.

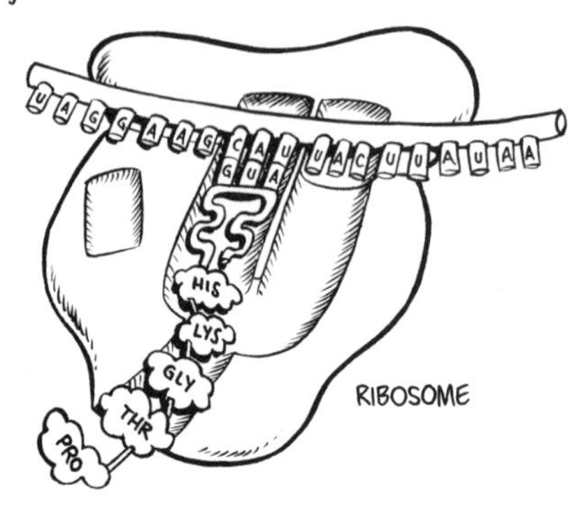

RIBOSOME

RNA: Ribonucleic acid. A single-stranded molecule that serves as the intermediary between DNA and protein production.

SEGREGATION: Separation of paired alleles on homologous chromosomes during meiosis, ensuring each gamete does not receive duplicate alleles.

STEM CELL: A cell that is not completely differentiated and retains the ability to produce daughter cells that can further differentiate.

STROMATOLITES: Fossilized layers of cyanobacteria, which are also known as blue-green algae.

TAY-SACHS DISEASE: An autosomal recessive disease ultimately resulting in neuronal death.

TELOMERASE: An enzyme responsible for the addition of telomeres to chromosome ends.

TELOMERE: A single-stranded, non-coding molecular cap on the ends of eukaryotic chromosomes that prevents joining with chromosome fragments.

TELOPHASE: The final stage of cell division, during which the chromosomes arrive at the poles of the cell and the cytoplasm divides.

THYMINE: Nucleic acid of DNA.

TOTIPOTENCE: The quality of a cell that allows it to retain the ability to differentiate into any cell type.

TRANSCRIPTION: The process by which mRNA is synthesized from a DNA template, resulting in the transfer of genetic information from the DNA molecule to the mRNA.

TRANSFER RNA (tRNA): RNA molecule responsible for transferring amino acids to the ribosomes for protein assembly.

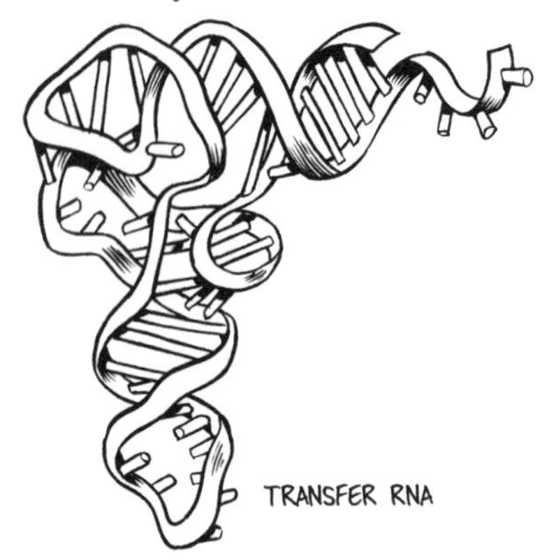
TRANSFER RNA

TRANSLATION: The process by which nucleic acids of DNA are recoded into amino acids for protein synthesis.

URACIL: Nucleic acid of RNA.

VECTOR: An organism or virus that transmits an infectious agent from one host organism to another.

VIRUS: Transmissible DNA or RNA enveloped by glycoprotein and requiring a host organism for replication.

ZYGOTE: The term applied to the diploid cell resulting from the union of a male and female gamete until the first cleavage.

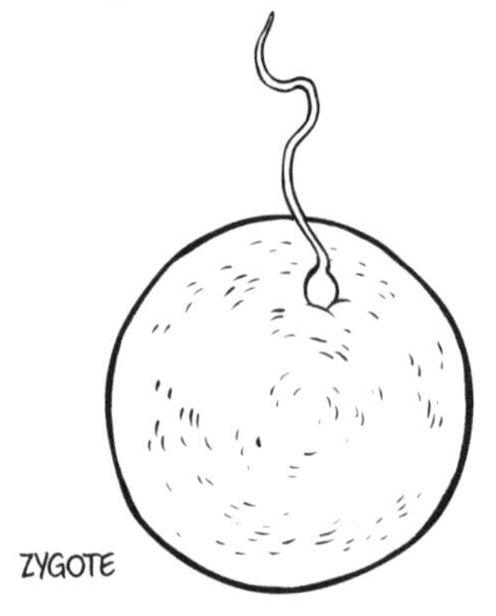
ZYGOTE

ABOUT THE SCRIPTER

MARK SCHULTZ has been cartooning, illustrating, and writing for longer than he cares to admit. His best-known creation is the award-winning speculative adventure comic book *Xenozoic Tales*, which ran for fourteen issues and has been adapted to television as the animated series *Cadillacs and Dinosaurs*. He has also created and written the undersea adventure *SubHuman* with paleontologist Michael Ryan. In addition to his own works, Schultz has drawn or scripted many popular fictional icons, including Superman, Flash Gordon, Tarzan, the Spirit, and Conan of Cimmeria. Currently, he writes the syndicated Sunday comic strip *Prince Valiant*, and continues producing material for his *Various Drawings* book collections. His most recently published work is the illustrated novella *Storms at Sea*. Schultz's scripts and illustrations have garnered five Harvey Awards, two Eisners, and an Inkpot. He lives in northeastern Pennsylvania.

ABOUT THE ILLUSTRATORS

Unrelated cartoonists ZANDER CANNON and KEVIN CANNON have been together as a studio since 2004, but their work in comics stretches back to 1993, with such titles as *The Replacement God* (Slave Labor Graphics and Image Comics, 1995), *Top Ten* (America's Best Comics, 1999), and *Smax* (America's Best Comics, 2003). They have also done various work for DC Comics, Dark Horse Comics, the National Oceanic and Atmospheric Administration, and the University of Nagoya. Their studio work includes the graphic novel *Bone Sharps, Cowboys, and Thunder Lizards* (G.T. Labs, 2005) and a number of other comics, as well as illustration, design, and animation work. They live and work in Minneapolis, Minnesota, and claim to be card-carrying members of the International Cartoonist Conspiracy.